The Indispensability Trap

Why Becoming Essential Caps the Fortune, From the Railroads to AI

Alan Shimel

The Indispensability Trap: Why Becoming Essential Caps the Fortune, From the Railroads to AI

Published by Fruition Harbor Press

Highland Beach, Florida

ISBN 979-8-9970373-0-7 (paperback)

ISBN 979-8-9970373-1-4 (hardcover)

ISBN 979-8-9970373-2-1 (e-book)

Library of Congress Control Number: 2026920066

For Bonnie, who has lived life's every chapter with me, including the ones that were hardest to write.

And for Landon and Bradley, who will always be the best thing I ever helped build.

How do we know when irrational exuberance has unduly escalated asset values, which then become subject to unexpected and prolonged contractions?

—Alan Greenspan, Chairman of the Federal Reserve, remarks at the American Enterprise Institute, December 5, 1996

Contents

Foreword

by Daniel Newman, CEO and Co-Founder, The Futurum Group

I am about as bullish on artificial intelligence as a person can be and still be taken seriously by a public market audience. I have said on national television, more than once and without hedging, that this is the largest technology-driven value creation cycle of our lifetimes. I believe that. I have not changed my mind while reading this book.

Which is a strange thing to write at the front of a book that spends nineteen chapters explaining why the people funding that cycle may not keep the fortune they are creating.

That tension is the reason I agreed to write this. Alan Shimel has produced the most useful kind of book about AI: One that does not tell you what to think about the outcome but shows you exactly which question to ask. And the question he asks is the one almost nobody in the capex debate is asking. Not "is the demand real." Not "will the models get better." Both of those have obvious answers, and both are yes. His question is harder. When this much capital pours into a single build-out, who ends up holding the value it creates?

Four times before, the answer has been: Not the people who built it.

The Man Asking Has the Receipts

Most of what is written about this moment is written by people pattern-matching from outside it. Alan is not doing that. He was in the room.

The book opens in Houston in the spring of 2000, with a young executive at an ASP called Interliant flying down to unwind a bandwidth contract that went underwater in under a year, and discovering that the counterparty was Enron, staffed with people who could not explain what a DS3 line was. That is not a metaphor Alan found in a research report. That was his Tuesday. He lived the fourth build-out, took losses in it, and then spent the next twenty-five years building the media and analyst infrastructure that covers the fifth.

I have had the good fortune to work alongside Alan since Techstrong became part of Futurum, and I will tell you what I have learned watching him operate. He is congenitally allergic to the press release. He has seen enough cycles to know that the loudest number in a quarter is usually not the important one. That instinct is all over this book.

The Thesis, Stated Plainly

Every foundational technology splits into two layers. There is a grid, which is capital-hungry, physically enormous, and essential. And there is an appliance layer built on top of it, which is where the durable money ends up.

Twenty-five years after the golden spike, a quarter of American railroad mileage sat in receivership. The fortunes went to Carnegie's steel, to Rockefeller's rebates, to Sears and Montgomery Ward, to Chicago. The Bell System built the most magnificent innovation engine in the history of business, and the government took most of it away in 1956, then took the rest in 1984. Insull electrified urban America and died a public villain while the appliance makers compounded. The fiber glut of 2001 destroyed the people who owned the pipes and handed Google, Microsoft, and Amazon a paved road.

Four for four. That is the Indispensability Trap. Becoming essential is not the same as becoming rich, and in these cycles, it has repeatedly been the opposite.

I want to be clear that I do not think Alan is wrong about any of that. I think the historical record is the historical record.

Where I Agree, Without Qualification

Commoditization at the grid layer is not a risk to be managed. It is a schedule.

Anyone modeling today's inference margins out five years has not read Chapter 10 of this book. The open-weight wave is real, model quality is converging faster than any forecast anticipated, and pricing is collapsing in near-lockstep. I have argued this on air and in print: The assumption that frontier capability stays scarce and premium-priced is the single most fragile assumption in the entire AI investment thesis. Alan documents why. The chapter on China's two loops should be required reading for anyone building a five-year model on a Western pricing curve.

Where I Read the Same Data Differently

Here is my honest reframe, and I offer it as a friendly amendment rather than a rebuttal.

The Indispensability Trap is a book about distribution, not creation. Every one of the four prior build-outs made the world structurally, permanently richer. The railroads collapsed and America got a continental market. The utilities collapsed and America got an appliance economy. The fiber players collapsed and we got the cloud, which is now a trillion-dollar-plus industry that did not exist. The trap describes who keeps the fortune. It does not describe whether the fortune exists. It always has.

Cheap distance built Sears. Cheap current built General Electric and everything downstream of a wall outlet. Cheap intelligence builds something we have not met yet, and I would rather be early to that than correct about the margin structure of the people laying the cable.

Three things I would add to the ledger.

First, funding structure matters. This build-out is being financed primarily out of the operating cash flow of the most profitable enterprises ever assembled, not out of speculative debt against demand that does not yet exist. That is a materially different balance sheet than 1999. It is not immunity. Chapter 8's mapping of vendor financing and circular deal structures is exactly the right place to look for where that argument breaks.

Second, we have already entered what I have been calling the Hard ROI Era. Through 2025 and into 2026, the market stopped paying for narrative and started demanding proof. Boards are asking for payback periods. That discipline arrived faster in this cycle than in any of the other four Alan chronicles, and it is the healthiest development in the sector.

Third, the winners of this cycle will not be sorted by who sells AI. They will be sorted by who plays offense with it. The companies restructuring their own cost base and revenue model around AI are compounding right now, quietly, while the argument about the vendors continues.

What I Do Not Have a Good Answer For

Serious optimism means owning the parts that are unresolved.

Power and interconnect timelines are the real constraint, and they are measured in years we do not have. Depreciation schedules on accelerated compute deserve far more scrutiny than they receive. Chapter 13 introduces a variable almost nobody has priced, which is a state that can now switch a model off. And Chapter 17 documents the constituencies paying real costs in water, land, grid capacity, and displaced work while holding no equity in the outcome. I do not think that is sustainable, and I do not think the industry has an answer yet.

Read This as an Operator

This is not a bear book, and anyone who reduces it to one has misread it. Alan says it himself in the conclusion: He is bullish on AI without hedges, and he believes the trap is real. I hold exactly the same two positions.

The people building the fifth grid are reaching for something extraordinary, and much of what they build will outlast the fortunes made and lost building it. That is the pattern, and it is a good pattern for the world.

But do not confuse being indispensable with being safe. Read the history first. Then place your bet.

Daniel Newman

Austin, Texas

Prologue

The Room in Houston

In the spring of 2000, I flew to Houston to get the company I had helped build get out of a bad contract I had helped get signed.

I was on the executive team of Interliant, an application service provider that had gone public the summer before. On paper, the stock and options I was sitting on made me a real, live internet millionaire. Of course, paper is just paper.

The contract at question was for bandwidth, one of our biggest costs. A year earlier I had helped lock in what we thought was a great deal with Bernie Ebber's MCI WorldCom. We could buy as much bandwidth connectivity as we could use and more. We felt good about it. Costs locked in, we could plan around it.

Within a year, the street price of that same bandwidth had collapsed by nearly half, and it was still falling. A dozen companies were racing to bury fiber-optic cable across the country faster than anyone could possibly use them. So when a company got in touch to say they were setting up a commodity-style trading operation for bandwidth, one that could swap our high-priced contract for street pricing, our CEO, COO, and I got on a plane.

They put us up in a conference room at the top of a glass tower in downtown Houston. Around a large oval table sat eight or nine of their people, and my CEO noticed the thing that should have told us everything before a word was spoken: No two of them were from the same division. Trading. Broadband. Network. Finance. Another from a department none of us could quite identify. They had been assembled to explain a deal that made less sense the longer they explained it, something about buying our bandwidth from us and then selling it back to us, repriced along the way. They explained it with total confidence and complete fluency.

The company was Enron.

We didn't do the deal, which is one of the better decisions I have been part of. Within two years Enron was a synonym for fraud, WorldCom had collapsed in the largest bankruptcy in American history to that point, the fiber that a dozen companies had raced to bury was trading for pennies on the dollar, and the stock and options that had made me a paper millionaire were paper again. The bandwidth kept getting cheaper the whole way down. The thing that took me twenty years to fully understand was this: The collapse wasn't the end of the story. It was the middle. The cheap bandwidth that ruined the men who built the fiber became the free raw material for everything that came after, the web, the cloud, the streaming economy, a generation of companies built on top of a network they never had to pay full price for.

I didn't know it that afternoon in Houston, but I was watching the fourth great build-out learn a lesson that three generations of infrastructure builders had learned before it. Twenty-five years later, the fifth build-out is spending ten times the money, and the question this book exists to ask is whether anyone in the room has done the reading.

Part One: The Pattern

Two chapters that name the trap and set the terms for everything after.

Chapter 1

The Trillion-Dollar Bet

> Financial capital…takes the economy on a frenzied ride up a paper-wealth bubble.
>
> —Carlota Perez, Technological Revolutions and Financial Capital, 2002

Drive far enough into Richland Parish in northeastern Louisiana, and it's like driving back in time. In this almost forgotten corner of the Old South, you go back to a period when cotton was king, when sharecroppers—bent over for hours with huge sacks strung over their backs—would be picking the puffs from the spiny plants until their fingers bled. Yes, Richland had its moment with King Cotton, but that was a long time ago.

Now you come to a place where one of the largest companies on earth is preparing to spend, by Bloomberg's reporting, well over $200 billion. The project is called Hyperion, it belongs to Meta, and when it is finished it will draw about five gigawatts of electricity, roughly what New York City pulls from the grid at any given moment, day and night, year round. Feeding it will take ten new gas-fired power plants, thousands of acres, and a rework of the region's water and transmission that the parish could not have imagined a decade ago.

To land it Louisiana went back to the Kingfish days and handed Meta property tax reductions approaching 80 percent, modified existing legislation, and put state officials under nondisclosure agreements. It is, by any measure, one of the most ambitious infrastructure projects in the history of the American South.

Next to all those big numbers, one metric sticks out like a sore thumb. When Hyperion reaches its long-term stride, sometime around 2035, Meta expects it to employ about five hundred people. Two hundred billion dollars of investment for five hundred jobs? Not exactly the return Richland was hoping for.

That is not a criticism of Meta, and it is not an accident. It is the shape of the thing. For most of the twentieth century, towns like the ones in Richland Parish chased factories—a Toyota plant, a Caterpillar line, a John Deere works—because a factory meant thousands of jobs and a future. Hyperion, on the other hand, consumes power on the scale of a great city, land on the scale of a manufacturing complex, and will employ about as many people as a large high school. The resources are industrial. The employment is not. Richland wasn't lucky enough to land any of those previous factory projects, so when Meta came calling, it was on par with offering water to people in the desert. It was the only offer they had, so they took it.

Richland Parish is not unusual in this. Outside the town of Broadview, Montana, a gigawatt-scale campus is rising whose permanent workforce, when the construction crews are gone, could be seated in the town's only restaurant on a slow night. One planned Utah project pencils out to somewhere between $35 and $50 billion of investment to start, against, by the count of the local and state government officials pushing hard to land it, between one and two thousand permanent jobs. This is the ratio repeated across the country. Capital on the scale of the steel age; employment on the scale of a corner store. The deeper you go into artificial intelligence, the less it looks like software and the more it looks like heavy industry. Well, heavy industry with a very peculiar ratio of capital consumed to people employed.

Before I go any further, know this. I run a media network that covers this industry, and I earn a fair share of my living being paid by the companies building these projects: the cloud providers and chipmakers, the infrastructure vendors filling the paragraphs above. This has afforded me a ringside seat and a bit of an inside view. The argument I am going to make is not especially good news for a number of these companies. It is certainly not the argument a person in my position would make if their goal was to keep every sponsor comfortable. You should read what follows knowing who is writing it and who pays his bills. Then you should judge the argument on whether it is right.

My argument is this: We are watching the largest infrastructure build-out of our lifetimes, and almost everyone is watching the wrong part of it. The attention, the awe, the money, and the fear are all pointed at the grid itself, at the data centers, the power plants, and the chips, at the companies laying down the physical substrate of machine intelligence. That is where the spectacle is. It is not, if history is any guide, where the lasting fortunes will end up.

When a foundational technology arrives, the durable wealth it creates rarely stays with the people who own the core infrastructure. It migrates. It goes to the people who build on top of that infrastructure once it becomes cheap, ordinary, and everywhere. This is like the way businesses now build on the assumption of electricity without a thought for building the power company. The builders of the grid get the spectacle and the capital risk at the front end, but the builders on top of it get the wealth of the century.

This is not a hunch about the future. It is a pattern from the past, and it has run at least four times before.

Consider the first of these, the railroads, because it is the clearest. In the second half of the nineteenth century, the United States laid down the greatest transportation network the world had ever seen. The men who celebrated at the driving of the golden spike believed they were building the defining industry of the age. It was a reasonable belief, but they were only half right. They were building the defining *infrastructure* of the age. The defining *industry* turned out to be something else.

The railroad companies themselves, taken as a group, were financial disasters. They were overbuilt, looted by their own promoters, driven into bankruptcy in waves, and finally handed to a federal commission that told them what they were allowed to charge. Meanwhile the greatest fortunes of the era were built by people who did not own a mile of track. Andrew Carnegie sold the railroads their steel. John D. Rockefeller shipped his oil over them and bent their freight rates into a weapon. Richard Sears and Montgomery Ward turned the network into a store the size of the country. The value did not stay on the rails. It climbed off them, onto the layer above, into the hands of people who treated cheap transportation as a free input and built something on top of it.

Hold that shape in your mind, because you are going to see it again in several chapters of this book. Then you are going to see the pattern setting up to happen a fifth time, with intelligence in the role that transportation played then.

There is a name worth giving to what is being built right now, the old name, "data centers," badly undersells it. What Meta is pouring into Richland Parish, and what the whole industry is racing to erect across the country, is an "intelligence grid." A national layer of compute and power as foundational to the coming economy as the electrical grid was to the last one. I will use that term throughout this book because it carries the right freight. It puts the AI build-out in a line of descent alongside the rail network, the telephone network, the power grid, and the internet, each of which was, in its day, the indispensable grid that everything else ran on.

The scale of the fifth run is worth pausing on because it dwarfs everything that came before. The four biggest hyperscalers alone are on course to spend around $725 billion this year, up more than 70 percent from the year before. Goldman Sachs puts the cumulative bill for this build-out at roughly $7.6 trillion between now and 2031.

Precise figures in this book will always be a little out of date by the time you read them, and, honestly, nobody knows the final number; what matters is the order of magnitude. And it is not just the famous names. Behind the hyperscalers is a whole second tier—the neoclouds, the Oracles, and the CoreWeaves, the colocation firms, and the sovereign builds—all pouring concrete of their own. The International Energy Agency expects the electricity consumption of all of this to roughly double by 2030 to something on the order of Japan's entire annual usage. The grid going up is wider and more crowded than the headlines suggest, and that breadth does not weaken the argument. It sharpens it.

Underneath that pattern is machinery, a repeatable process with three moving parts. The first is commoditization, which erodes what the grid's owner can charge from below. The second is a paradox, which guarantees that as the grid gets cheaper the world uses dramatically more of it. The third is the trap this book is named for, the Indispensability Trap, which caps the essential from above. Running through all of it is a rhythm, two acts—one of frenzy and the other of deployment—that every one of these build-outs has followed. The next chapter lays the machinery out plainly and tells you exactly how I intend to keep myself honest about it. A pattern this convenient deserves your suspicion before it earns your agreement. For now it is enough to hold the pattern: The grid gets built, the grid gets cheap, and the fortune climbs off it.

There is another thing to say before we go back to the nineteenth century, and it concerns the loudest voices in the room. The argument of this book runs directly against some of them. There is a serious, well-funded, and genuinely intelligent case being made that this time infrastructure will win. That the companies pouring capital into the grid are buying the most valuable real estate of the coming decades. That the factories are not overbuilt but, in fact, underbuilt, and that the returns will go vertical as demand explodes.

I take that case seriously, and I will give it a full and fair hearing. The people making it are not fools, and they may turn out to be right. But I think they are making, in a new costume, the same assumption the railroad men made when they looked out over their track and saw the future. They are confusing the infrastructure with the industry. They think they are in the railroad business when they are actually in the transportation business.

Whether the people building the intelligence grid have understood that distinction any better than the railroad men did is the question this book exists to ask. The five hundred jobs in Richland Parish, sitting next to a capital bill that runs well into the hundreds of billions of dollars, are the first clue that something in this story does not add up. The rest of the clues are a hundred and fifty years old, and they start with a golden spike in the Utah desert.

Chapter 2

Grid and Appliance

> History never repeats itself, but the Kaleidoscopic combinations of the pictured present often seem to be constructed out of the broken fragments of antique legends.
>
> —Mark Twain, *The Gilded Age*, 1873

Before we go to Utah, or anywhere for that matter, I want to put the strongest objection to my own thesis on the table: A pattern drawn through four examples is not a law but a highlight reel. Give a clever writer the run of two centuries of economic history and the freedom to pick his cases, and he can draw a confident line through almost anything. Four build-outs in which the people who owned the infrastructure lost and the people who built on top of it won, chosen by the man making the argument, proves very little on its own.

Maybe I went looking for the railroads and the telephone and electrification and the internet precisely because they fit. Maybe I quietly stepped around the build-outs that didn't. Maybe the pattern is real but rare, and I am dressing up a coincidence as a law because I have a book to sell and an AI boom to explain. That is a fair thing to suspect. I could ignore it and say, well, this is what I believe. The intellectually responsible thing to do, though, is not to wave it away but to answer it, which is what this short chapter is for.

The answer is that this is not four anecdotes. It is one mechanism observed four times, and a mechanism is a very different thing from a coincidence, because a mechanism makes predictions. If I can describe how the value moves, step by step, in a way that does not depend on which technology we are talking about. If that same machinery turns out to be running underneath the railroads and the telephone network and the power grid and the internet, then the pattern is not something I selected. It is something I found. A mechanism that has run four times under four completely different sets of technical and political circumstances is a mechanism you should expect to run a fifth time.

So let me describe the machinery plainly, because the rest of the book is going to assume you can see it working.

The Two Layers

Every foundational technology, once it matures, splits into two layers, and almost all of the confusion about who wins and who loses comes from not keeping them separate.

The first layer is the infrastructure itself, the thing that gets built. The rails and the roadbed. The copper wire and the switching network. The generators and the transmission lines. The fiber and the data centers. This is the layer that requires enormous capital to construct, that inspires the awe and the magazine covers and the trillion-dollar forecasts. This first layer looks, at least while it is being built, like the obvious place the money will end up. Call it the grid.

The second layer is everything built on top of the grid, everything that treats the grid as a given and does something valuable with it. The steel mill and the mail-order catalog that assumed cheap rail freight. The call center and the software company that assumed cheap telephony. The assembly line and the appliances that assumed cheap power. The search engine and the streaming service that assumed cheap bandwidth. This layer does not build the infrastructure; it consumes it, the way you consume electricity from a wall socket without a thought for the power plant. Call it the appliance.

The argument of this book, stated at the level of these two layers, is simple. The durable fortune ends up in the appliance layer, not the grid layer, and it gets there by a process you can watch happen.

The process has three moving parts: The first is commoditization. A grid, to be useful, has to become cheap and universal, and the moment it does, its owner loses the ability to charge what the thing is worth. A commodity by definition sells at the cost of producing it and no more.

The second is Jevons paradox, which guarantees that as the grid gets cheaper, the world does not use less of it but dramatically more. All of that new demand becomes raw material for the appliance layer to build on.

The third is the Indispensability Trap. The more indispensable the grid becomes, the more certainly society moves to cap its owner's returns and guarantee everyone access. The cap arrives through a rate commission, an antitrust suit, or an outright state takeover. Commoditization erodes the grid owner's pricing from below; the trap caps it from above. And the demand the grid unleashes flows past the owner entirely, up to the builders who assumed the grid was free.

That is the whole machine. Value does not vanish. It migrates, every time, from the grid to the appliance by those three forces working together.

The Company That Owned Both

There is one complication that has to go right here at the start. Without it the model is too crude, and a careful reader will break it in the first chapter of history.

The complication is that the grid and the appliance do not have to be owned by different companies. Usually they are, which is why the pattern is usually easy to see: the railroad went bankrupt while Carnegie got rich, two different balance sheets, a clean migration. But occasionally a single company is smart enough, or ruthless enough, to own the grid and climb into the appliance layer above it and capture both. The archetype has a name, and it is General Electric.

We will meet GE properly in the electrification chapter, so I will be brief. GE did not settle for building the dynamos, the grid-layer business that got regulated into a capped utility. It also built the motors and the light bulbs and the appliances, the layer where the value was migrating. By owning both it became one of the largest enterprises in the history of the world, while the pure-play power utilities beside it were turned into regulated plumbing.

GE is the proof that the thesis is not "infrastructure owners always lose." That is the crude version, and it is wrong. The precise version is this: Value migrates up from the grid to the appliance, relentlessly, and the only grid owner who keeps the fortune is the one who climbs up into the appliance layer along with it. The pure play gets stranded at the bottom and commoditized. The integrated giant who follows the value upward becomes GE.

Remember GE, because it is going to matter enormously at the end of the book. The companies building the intelligence grid today are, many of them, explicitly trying to be GE, trying to own the infrastructure and every layer above it at once. Whether they can, and what happens if even the top of the stack commoditizes underneath them, is the question the second half of this book is pointed at.

The Test

Now how do you keep yourself honest about a pattern like this, so you are genuinely finding it rather than imposing it? You fix the questions in advance, and you ask the same ones of every case, in the same order, and you let the answers fall where they fall.

This book puts six questions to each of its build-outs. How did the technology cross from novelty to necessity? Who paid to build it, and did the builders keep the wealth? Did the thing they built become a commodity, and did society step in to regulate it? Where did the durable fortune actually land? What did people fear the technology would do, and were they right? And, when we reach artificial intelligence, the one that matters most, how much of the pattern holds this time, and where does it break?

The discipline is in asking the questions the same way every time, because questions asked consistently can come back the wrong way, and that is exactly what makes them worth asking. This is the difference between a law and a highlight reel. If the fourth question, where did the durable fortune land, had come back "with the infrastructure owner" in even one of these four cases, the thesis would be in serious trouble, and I would owe you an explanation.

The thesis is convincing precisely because it does not. The same test, applied honestly to four independent build-outs, separated by a century and running on completely different technologies and politics, keeps returning the same answer. Besides GE there is one other "almost" exception to this rule, Cornelius Vanderbilt. Vanderbilt made a genuine fortune in the railroad industry. But he made his money consolidating and financing the rails in the unregulated years rather than operating them. Even still, he topped out at a fraction of what Carnegie and Rockefeller built on top of the network he ran. The test survives its hardest case. That is the kind of thing that ought to move you, and the kind of thing cherry-picking cannot produce. You cannot cherry-pick a consistent answer to a question you fixed before you looked.

Where It Might Break

The last thing an honest version of this argument has to do is tell you where it might be wrong, so let me do that now rather than pretend the pattern is airtight.

There are build-outs that fit less cleanly than my four. Some technologies never split neatly into a grid and an appliance layer, and the model has less to say about them. Some infrastructure owners have held their pricing power longer than the thesis would predict, at least for a while, usually by owning something with no substitute at all. And the integrated giant, the GE case, is a real and recurring exception that the crude version of the argument misses, which is why I put it in the model from the start rather than letting it ambush us later. I am not claiming a law with no exceptions. I am claiming a strong and repeated tendency with a mechanism behind it, which is a more modest and more defensible thing. That is more than enough to make the pattern worth taking seriously when it points at the present.

And then there is the largest possibility of all, the one the entire second half of this book is built around. It is possible that artificial intelligence is the case that breaks the pattern. Every analogy in this book is drawn from a technology that was, in the end, passive, that sat there and moved goods or current or bits and did exactly what it was told.

Artificial intelligence may be different in kind rather than degree, and if it is, some of the machinery I just described may not run the way it ran for the railroads. I do not think the reader is well served by a writer who hides that possibility to protect his thesis. I am putting it here, at the front, as a live question rather than a settled one. The pattern has held four times. Whether it holds a fifth time, against a technology unlike any of the four, is genuinely open, and it is the most interesting question in the book precisely because I cannot promise you the answer.

That is the model and the method: a grid that commoditizes, an appliance layer that inherits the value, three forces that move it from one to the other, a GE exception built in from the start, six questions asked the same way every time to keep the whole thing honest, and an open admission of where it might fail. Now we can go watch it run, and the place it runs most clearly, the place the whole pattern shows itself for the first time in the fullest light, is a golden spike in the Utah desert in the spring of 1869.

Part Two: The Great Build-out

Four industries—rails, telephony, electrification, and the cloud—that ran the pattern before.

Chapter 3

In the Railroad Business, Not the Transportation Business

> They let others take customers away from them because they assumed themselves to be in the railroad business rather than in the transportation business.
>
> —Theodore Levitt, "Marketing Myopia," *Harvard Business Review*, 1960

On a spring afternoon in 1869, at a wind-scoured spot in the Utah Territory called Promontory Summit, a crowd gathered to watch a man hit a railroad spike with a silver hammer. The spike was gold, the hammer was wired to a telegraph line, and when it finally struck home, the news went out to both coasts in the same instant. The continent was stitched together. It deserved the ceremony. A trip that had taken six months by horse or oxen-drawn wagon and could still take six weeks by ship around Cape Horn, had just been cut to about a week. It could be taken in a heated car at a price an ordinary family could begin to imagine. By any reasonable measure, this was one of the great engineering achievements of the age, and the people on that platform knew it.

What none of them could see was that within twenty-five years a quarter of American railroad mileage would be sitting in the hands of bankruptcy receivers, more than forty thousand miles of track, including the very transcontinental lines whose completion they were celebrating that afternoon. The companies that built the most heroic infrastructure of the nineteenth century turned out, with striking consistency, to be mediocre businesses. They were overbuilt, over-leveraged, milked by their own promoters, and eventually told by Washington what they were allowed to charge. Promontory was the summit in more ways than the surveyors intended or could have imagined.

The wealth the railroads created was real, and it was enormous. It simply pooled somewhere other than the railroads. It collected in the steel mills that supplied the rails, in the oil company that learned to bend freight rates to its purposes, in the mail-order houses that turned a national network into a store the size of a country, in the packing plants that could ship a slaughtered steer a thousand miles, and in the city that made itself the knot every shipment had to pass through. The men who got rich off the railroads, for the most part, did not run railroads. They understood something the railroad men did not, which is roughly the thing that the economist and Harvard Business School professor, Theodore Levitt would put into words ninety years later. They understood that the railroads were not selling track; they were selling distance. And once distance got cheap, distance became the platform that other people built their own empires on.

The railroads are where this pattern appears first and in the clearest light, which is why we start here rather than with electricity. Almost nobody remembers the nineteenth century as anything but the railroad era, and that gap, between what an age is remembered for and where its money actually went, is the real story.

From Curiosity to Nervous System

It is easy to forget how fast this happened. In 1850 the country had roughly nine thousand miles of operating track, already about as much as the rest of the world put together. Twenty years and a civil war later, it had passed fifty thousand. By 1890 it had crossed a hundred and sixty thousand miles. It would push through two hundred thousand by the turn of the century, more than Britain, Germany, France, Russia, and India combined.

Inside a single working lifetime, the railroad went from a regional novelty that skittish investors treated as a gamble to the physical substrate the entire economy ran on. A farmer in Kansas, a foundry in Pittsburgh, a dry-goods store in a town that existed only because the line came through, none of them could function without it. That is what indispensable looks like when you watch it arrive in real time. As the rest of this chapter argues, becoming indispensable was the beginning of the railroads' problem rather than the end of it.

The clearest illustration of how deep the dependence ran is something most people never connect to the railroads at all: It is the clock on your wall. Into the 1880s time in America was a local matter. Every town set noon by its own sun. This meant that noon in Chicago was several minutes off from noon in St. Louis and further still from noon in Omaha. By one count there were more than fifty local time standards scattered across the country. That was a quaint inconvenience when the fastest thing moving between two towns was a horse. It became an operational nightmare, and a genuine safety problem once you were trying to run heavy trains toward each other on shared track according to a printed schedule.

The railroads fixed this themselves. At the General Time Convention that met at the Grand Pacific Hotel in Chicago on October 11, 1883, railroad officials adopted a plan for five standard time zones. There would be one Atlantic and four continental zones. They set a date to switch. At noon on November 18, 1883, railroad clocks across the country were reset, some of them jumping backward or forward in the middle of the day, an event the newspapers of the time called the "Day of Two Noons." Most towns simply went along, because the railroad was their lifeline to the rest of the world and arguing with the schedule was pointless. Not everyone was thrilled about a private industry reorganizing something as elemental as time.

Three days after the reset, the *Indianapolis Daily Sentinel* complained that "the sun is no longer to boss the job." The *Sentinel* was right, and it lost. Standard time held. Congress did not get around to writing it into federal law for another thirty-five years, when Woodrow Wilson signed the Standard Time Act on March 19, 1918. By then the railroads had already imposed on the country a system it still lives by, which is about as pure a demonstration of infrastructure power as you will find.

There is a second, quieter consequence of that scale. To run an enterprise stretched across thousands of miles, with tens of thousands of employees and equipment that could kill people if mismanaged, the railroads had to invent the modern managerial corporation more or less from scratch. Alfred Chandler built much of his career documenting this, the shift from the owner-operator counting their own cash to the layered hierarchy of salaried professional managers coordinating a system too large for any one person to hold in their head. The railroads were the first organizations that had to solve that problem. The management playbook the rest of American industry would use for the next century was written, in large part, to run trains.

True to Jevons paradox, making transport cheap did not satisfy a fixed appetite for shipping and then stop; it multiplied the appetite, because a merchant who would never have shipped a plow across three states at the old prices shipped ten of them at the new ones.

A bit more background on Jevons and his paradox. An English economist named William Stanley Jevons had noticed the same pattern in coal a few years earlier, in 1865: Make a resource cheaper to use and total consumption tends to rise rather than fall, because low prices open uses that were never worth it before. That flood of now-economic shipping is the raw material the next set of fortunes were built out of, and the principle behind it will come back at the end of this chapter and again in the ones that follow when the resource getting cheap is not distance but intelligence itself.

Where the Money Leaked Out

None of this was built on private capital alone, and the way it was financed is where the trouble starts.

The federal government wanted a transcontinental line badly enough to pay through the nose for it. Private investors, looking at seventeen hundred miles of desert and mountains with no customers along the way, correctly judged it a terrible investment. Washington had to sweeten the deal. Under the Pacific Railway Act that President Lincoln signed on July 1, 1862, the railroads received government bonds worth between $16,000 and $48,000 for every mile of track they laid, scaled up for hard terrain, plus a land grant of 6,400 acres per mile, handed over in a checkerboard of alternating sections along the route.

Across all the federal acts between 1850 and 1871, Congress authorized something on the order of 175 million acres to the railroads, an area larger than Texas. The historian Paul Wallace Gates, working through the land records, found that roughly 130 million of those authorized acres were actually patented over to the companies, with the rest forfeited or never claimed. Even the smaller figure is close to a tenth of the United States. The theory was that the presence of the railroad would make the surrounding land valuable, the railroad would sell it to settlers, and everyone would come out ahead.

Now look at the incentive that structure created. The government paid by the mile. It did not pay for a well-engineered railroad, or a profitable one, or one routed sensibly. It paid for track laid down, so the rational move for a promoter was to lay a lot of track, bill as much as possible against it, and not worry so much about whether the resulting railroad was a going concern.

The men running the Union Pacific understood this perfectly. Led by Thomas Durant, the Union Pacific's vice president, and later run for maximum plunder by the Massachusetts congressman Oakes Ames and his brother Oliver, they set up a construction company with the impressive French name Crédit Mobilier. They had the railroad hire it to do the building, and then they charged the railroad wildly inflated prices, roughly $94 million for work that cost about $50 million. The difference flowed to the insiders, who happened to own both the railroad ordering the work and the company doing it. When the scheme needed political protection, discounted shares of Crédit Mobilier found their way into the hands of a good number of congressmen. The whole arrangement blew up into a public scandal in 1872 when the *New York Sun* broke the story in the middle of President Grant's re-election campaign.

Crédit Mobilier is usually filed under Gilded Age corruption, and it was that. It is more useful for our purposes as a clean illustration of a structural point. The fortunes being made in railroad construction were being made during the building, out of subsidies and contracts and stock, by people whose interest in the railroad as an operating business was close to zero. That is a different activity from running trains at a profit, and it is worth keeping the two separate, because the men who did the first thing often did very well while the enterprise that resulted from the second thing frequently did not.

The economist Carlota Perez, in her book *Technological Revolutions and Financial Capital: The Dynamics of Bubbles and Golden Ages*, has a useful pair of terms for this divide: financial capital, the money that chases a new technology for quick speculative gain, and production capital, the money that actually builds and runs the thing. Every technological revolution, in her telling, opens with the first kind in charge and only pays off once the second takes over.

The men looting the Crédit Mobilier contracts were financial capital in its purest form, trading on the excitement that an industry might exist rather than actually building one, and that split between the speculators and the builders is going to recur in every chapter that follows.

How badly did the operating businesses do? Twice within twenty years the overbuilding caught up with the whole sector and took the economy down with it. The Panic of 1873 and the far worse Panic of 1893 were, at their core, railroad crises set off by lines that had been built ahead of any demand that could pay for them. In the 1893 collapse roughly a quarter of all US rail mileage went into receivership, forty thousand miles and more, and the casualty list included the marquee names: the Union Pacific, the Northern Pacific, the Santa Fe. These were not fly-by-night operations. They were the transcontinentals, the crown jewels, and they went broke. What followed was consolidation, the bankrupt lines swallowed by stronger hands, until by 1906 seven groups controlled two-thirds of the mileage in the country. The romance of the open competitive railroad frontier ended in a handful of surviving giants and a lot of wiped-out bondholders.

To be fair to the other side of this story, there is another view. The "great giveaway" story about the land grants is overdrawn. The government loans were, for the most part, repaid. The railroads did not, on the whole, get fabulously rich reselling their land. And only a fraction of the country's mileage was actually built by land-grant lines. You can concede all of that and still lose nothing, because the point here is not that the subsidies were a swindle from top to bottom. The point is narrower and sturdier. The activity that reliably produced fortunes was financing and building the railroads, not operating them, and the operating companies, taken as a class, were chronically fragile businesses that periodically detonated.

The Bargain That Capped the Upside

Then society did to the railroads what it eventually does to any business that becomes genuinely essential. It took away their freedom to price.

The pressure came from below, from the people the railroads squeezed. A farmer who had no choice but to ship his grain on the one line that reached his town was at that line's mercy, and the railroads were not shy about it. They charged more where they had a monopoly than where they faced competition, gave secret rebates to their biggest customers, and routinely charged more for a short haul on a captive route than for a long haul on a contested one. The farmers organized, the Grange movement pushed state legislatures, and the states started passing laws to regulate rates. The Supreme Court initially let them, upholding state regulation of businesses "affected with a public interest" in *Munn v. Illinois* in 1877.

The states hit a wall in 1886. In the *Wabash* case, the Court ruled that a state could not regulate rates on interstate traffic because the Constitution reserves interstate commerce to Congress. That decision cut the ground out from under the state commissions and left a vacuum, and the political demand for control of the railroads was far too strong to leave it empty. Congress filled it the next year.

The Interstate Commerce Act, signed by President Cleveland on February 4, 1887, required that railroad rates be "reasonable and just," banned the rebates and the long-haul-short-haul games, and created the Interstate Commerce Commission to enforce it. The railroads became the first industry in America placed under federal regulation, and the ICC became the first independent federal regulatory agency of its kind, the template that the Federal Trade Commission, the Securities and Exchange Commission, and a long line of others would later copy.

Trace this arc, because it is the same arc that almost every chapter comes back to. A technology arrives and looks like a license to print money for whoever controls it. It becomes essential. The moment it becomes essential is the moment the public stops treating it as a miracle to be priced at whatever the market will bear and starts treating it as a common good that has to be affordable, fair, and available to everyone. Two things happen at that moment, and they are really one thing: Society moves to guarantee that the essential good reaches everyone, and it moves to cap what the owners can charge for it, because no one gets to promise universal access and keep unlimited pricing power in the same breath. The stronger the dependence, the harder both levers get pulled. The owners of the infrastructure do not necessarily go broke, though plenty of them did.

What happens is subtler and more permanent: Their upside gets capped, on purpose, by a society that has decided the thing they own is too important to be left entirely to them. It is the closest thing this book has to a law, and some version of it runs through every case ahead. Hold this thought for the next chapter, because a generation later a utility executive named Samuel Insull will look at the electric power business and, rather than fight this logic, walk straight into it on purpose. But that is the next chapter's story. The railroads got there first, and mostly by force.

Where the Fortune Actually Went

So if the railroad operators were, as a group, running hard just to stay in place and then handing their pricing power to a federal commission, where did the era's real money end up?

Some of it did end up with railroad men, the cleanest exception to everything I have argued so far is Cornelius Vanderbilt, the Commodore, who died in 1877. At the time his fortune was estimated at something like a $105 million, about one eighty-seventh of the entire national economy at the time, which made him the richest man in America. That is a railroad fortune, and there is no other way to state it. But look at how he built it, because the details matter.

Vanderbilt made his money in shipping first, then by consolidating the fractured lines between New York and Chicago into a single system, squeezing rivals in rate wars and playing the stock of the railroads he was fighting over. He got rich owning and combining and financing railroads in the years before the regulators showed up, not by patiently operating a railroad at an honest margin over decades.

The Central Pacific's "Big Four" and the market operators like Jay Gould worked variations of the same game. The pattern holds even for the winners: The railroad fortunes were consolidation-and-finance fortunes, made in a specific window, and the underlying operating business kept demonstrating how weak it was by going bankrupt.

Now set Vanderbilt next to the two men who ended up bigger than any railroad operator and notice what business they were actually in.

Andrew Carnegie sold the railroads their steel, and he chose that position deliberately; the choice is a lesson in and of itself. Carnegie knew the railroads from the inside. He had started as a boy delivering telegrams, taught himself the wire well enough to be hired by Thomas Scott of the Pennsylvania Railroad, one of the best-run lines in the country. By the age of twenty-four, he had risen to superintendent of its western division, close enough to the top to watch how the business worked and where its money actually went. From Scott he learned the discipline that would make him rich, an obsession with knowing to the penny what it cost to move a ton or roll a rail.

A man with that vantage could have spent his life climbing the operating side of a great trunk line. Carnegie looked at it and decided he would rather supply the railroads than be one. He saw that the country was going to lay track and, later, raise steel-framed buildings faster than anyone could imagine, and that whoever produced the rails and beams to do it would sell everything he could make to a market that could not get enough.

Carnegie built the most efficient steel operation in the world and pointed it straight at that demand. He was ruthless about cost, obsessive about the newest process, willing to scrap a nearly new mill for a slightly better one, and the result was a company that could undersell everyone and still make money in the depths of a depression. He was not running trains. He was arming the people who did, and the arms business turned out to be the better business.

In 1901 J.P. Morgan bought him out to assemble a steel trust, and the two men settled the price of Carnegie Steel on a single slip of paper, roughly $480 million, with Carnegie's personal share coming to better than $225 million, paid out in fifty-year gold bonds so bulky he had a vault built across the ocean to hold them.

The company Morgan built around that purchase, US Steel, was capitalized at about $1.4 billion, the first billion-dollar corporation in the history of the world and a figure equal to something like 4 percent of American GDP at the time. The deal made Carnegie, for a moment, the richest man alive. When they shook on it, Morgan reportedly told him, "Mr. Carnegie, I want to congratulate you on being the richest man in the world." He got there by selling to the railroads rather than being one, and he got there on purpose, having watched the operating business up close and chosen the layer above it.

John D. Rockefeller did something subtler than Carnegie and, in the end, even bigger. He did not supply the railroads or ride them so much as turn them into an instrument. He understood a weakness in the business we have already seen: The railroads were desperate for volume, drowning in fixed costs, terrified of running half-empty trains, and therefore willing to make quiet deals with any shipper big enough to fill their cars. Rockefeller made himself that shipper. He consolidated the refiners of Cleveland until Standard Oil moved enough oil to matter to any railroad that wanted the traffic, and then he used that leverage to extract terms no one else could get.

The terms were extraordinary, and they show exactly how the value moved. Standard Oil negotiated secret rebates, a kickback on its own freight, so that the posted rate everyone paid was, for Rockefeller, quietly discounted. That alone would have been a decisive edge, but he went further and secured drawbacks, by which the railroads paid him a cut of the freight his competitors shipped. A rival refiner would pay the railroad to haul his oil, and a slice of that payment would flow to Rockefeller, who was thereby paid by his competitors for the privilege of competing with him. The most audacious version of this, the South Improvement Company scheme of 1872, was so brazen it collapsed under public fury before it fully launched, but the underlying pattern survived and did its work. The railroad hauled the oil and collected the freight. Rockefeller collected the margin, and then the industry.

By 1913 his stake in the successor companies of Standard Oil was worth about $900 million, on the order of 2.5 percent of the entire national economy; three years later, in September 1916, the papers announced that he had become the first billionaire in American history. He owned no significant railroad. He had simply figured out that the party who controls the flow across a network can be worth more than the party who owns the network, and he proved it at a scale no railroad president ever matched.

Then there is the layer that looks the most like a modern software business, the mail-order houses, and they are the purest illustration in this whole chapter of building an application on top of somebody else's network. In 1872 a traveling dry-goods salesman named Aaron Montgomery Ward, who had watched rural customers get gouged by the only general store within a day's ride, printed a single sheet of paper listing about a hundred and sixty items at prices a city buyer would pay, and mailed it to farmers who could order by post and receive by rail. The idea was to cut out the middleman and let the network do the reaching. It worked well enough that within two decades the single sheet had become a catalog running to hundreds of pages and sales running to millions a year.

Richard Sears, a railroad station agent who started by selling watches to pass the time, followed with a partner named Alvah Roebuck. And Sears and Roebuck outran even Ward. Under Julius Rosenwald, who took over the merchandising, Sears grew from under $15 million in annual sales at the turn of the century to $50 million by 1907, and the catalog swelled into the object rural America called the Wish Book, a thousand pages offering everything from plows to violins to entire houses shipped in numbered pieces.

None of these firms owned a foot of track. What they owned was a catalog, a mail-order plant in Chicago engineered to fill an order in hours, and a business model that assumed the existence of cheap, reliable, nationwide shipping the way a fish assumes water. They ran on the railroads, plus two pieces of public plumbing that arrived to complete the system: Rural Free Delivery in 1896, which carried the catalog and the mail order out to the farm gate, and Parcel Post in 1913, which carried the package back.

A family in rural Nebraska could open the book at the kitchen table, choose from thousands of goods once available only in a distant city, and have them turn up at the local siding within days, and at prices the local store could not touch. Rosenwald grew wealthy enough on that arrangement to give much of it away, funding, in partnership with Booker T. Washington, the construction of 4,977 schools for Black children across the segregated South between 1912 and 1932. He built one of the great fortunes of the age without laying a single rail, by owning the catalog and the customer and letting the railroads carry the weight.

The same shape shows up in meatpacking. The story of Gustavus Swift is worth telling in full because he had to fight the railroads to win, which sharpens our point rather than blurs it. Before Swift, cattle traveled to eastern markets alive in stock cars, losing weight and sometimes dying on the way, so that a buyer in New York was paying rail freight on hooves, hide, and bone he would only throw away. Swift saw that if you slaughtered and dressed the animal in Chicago and shipped only the salable meat, you could serve the whole country from one efficient center, provided you could keep the meat cold across a thousand miles in summer.

He hired an engineer named Andrew Chase to design a practical ice-cooled car and then ran into a wall: The railroads did not want it. They had sunk capital into stock cars and animal pens and the whole live-cattle traffic, and dressed beef moving cold threatened all of it, so they not only refused to build the cars but banded together and set punishing rates on whatever dressed meat they would agree to carry. Swift built the cars himself and routed his traffic over the one line with no stake in live cattle, the Canadian-owned Grand Trunk Railway, which ran him through Michigan and Canada to the eastern markets.

Within a year of the first cars rolling in 1880, he had nearly two hundred of them moving some three thousand carcasses a week to Boston. He had reorganized an entire industry around cheap long-haul transport. Swift and his rival, Armour, became giants of the age. The railroads carried the beef. The packers who rebuilt the business around the network kept the fortune, and they kept it partly by beating the railroads in a fight the railroads started.

The pattern shows up not just in companies, but in the geography itself. A single city, Chicago, became the richest place in the interior of the continent not because it grew or made the most, but because it sat at the knot where the rail lines of the West met the rail lines of the East, and everything moving between them had to pass through and be handled.

The institutions the city invented to handle that flow were themselves the money machines: the grain elevators that graded and stored the harvest, the futures market that let people trade the grain before it arrived, the great stockyards that concentrated the nation's meat in one place. None of these produced a bushel or raised a steer. They monetized the passage of goods across the junction. William Cronon's history of the city, *Nature's Metropolis*, is, read a certain way, four hundred pages of a single argument, that enormous value accrued to whoever controlled the point where the network converged, rather than to whoever owned the line that reached it.

Put the numbers side by side and the migration is not a metaphor; it is a measurement. The greatest railroad operator, Vanderbilt, topped out around 1 percent of the economy, and he did it through consolidation in the unregulated years. The supplier who armed the rails, Carnegie, and the shipper who rode them, Rockefeller, each ended up with something closer to 2 and 2.5 percent respectively. Fortunes roughly double the biggest operator's, built beside and on top of the network rather than inside it. Meanwhile the network itself, the thing everyone was watching, was going through receivership in waves and being handed to a federal commission for pricing. The money did not stay with the grid. It climbed.

The Fear They Got Right

Every one of these build-outs frightened the public in some particular way, and which fear takes hold tells you a lot about what regulation comes next. The railroads did not scare people the way electricity would later scare them, with stories of a current that could kill you in your kitchen. The railroads scared people as a concentration of power. The image that stuck was the octopus, the railroad as a many-armed monopoly with a grip on the farmer, the merchant, the small town, and the legislature all at once. Frank Norris built a whole novel, *The Octopus*, around exactly that picture at the turn of the century.

The fear was substantially correct. The railroads really did discriminate on rates, really did buy legislators, really did use a chokehold on transport to squeeze everyone downstream of them. The octopus was not paranoia. It was reporting. And because the fear was grounded in something real, the regulation that answered it stuck, and hardened, and stayed on the books in one form or another for most of a century, until deregulation finally arrived with the Staggers Act in 1980 for an industry that, by then, was a shadow of the menace it had once been.

That distinction, between a fear of what a technology's owners will do with their power and a fear of what the technology itself might do, is going to matter enormously later in this book. The railroads, the telephone, electricity, the internet, every one of them provoked the first kind of fear, the fear of concentrated economic power, and every one of them got answered with some mix of antitrust and utility regulation.

Artificial intelligence is shaping up to be the first of these infrastructures to provoke the second kind of fear as well, a fear about the technology's behavior rather than its owners' greed. That is a genuinely different problem, and it is going to demand a genuinely different kind of answer. I am flagging it here and leaving it, because the railroads have nothing to teach us about that second fear. They do have a great deal to teach us about the first.

What the Railroad Men Could Not See

That brings us back to Levitt, and to the line that gives this chapter its title.

Writing in 1960 and looking back at an industry that had spent decades sliding into irrelevance, Theodore Levitt argued that the railroads' real mistake was one of self-definition. The demand for moving people and freight had not shrunk. It had grown enormously. The railroads lost anyway, Levitt said, because they thought they were in the railroad business rather than the transportation business, and so when the trucks and the cars and the planes and the pipelines showed up, the railroad men saw competitors to be resented rather than a business they were already in and could have owned. They defined themselves by the thing they built instead of the need they served, and the need walked off to whoever served it better.

Levitt's example is not airtight. Later critics have pointed out that the railroads' decline was not purely a failure of imagination. Real substitutes genuinely did arrive, the interstate highway system was a massive public subsidy to their competition, and, most relevant to everything this chapter has argued, the railroads were by then wrapped so tightly in rate regulation that they could not have priced or diversified like a free "transportation" company even if every executive had read Levitt in advance. The regulatory cage is part of why the railroads stayed railroads. They were not entirely free to become anything else. That does not break Levitt's point so much as complete it. The category error was real, and it was reinforced by a regulatory structure that had frozen the industry into a single definition of itself.

For now hold the whole shape of it in your head, because the reason to spend a chapter on nineteenth-century railroads in a book about artificial intelligence is that the shape is about to rhyme.

Start with Levitt's category error, because it is the one already happening. The companies pouring capital into frontier models right now are at risk of believing they are in the model business, the way the railroads believed they were in the railroad business. The durable position is almost certainly the larger category the models merely serve, the way transportation was the larger category the railroads served. Intelligence, not models, the way it was transportation, not railroads.

Then take the picks-and-shovels wrinkle. Most people get it backward when they look at today's market and conclude that the infrastructure owners have already won. Enormous fortunes were made during the railroad build-out. They were made by the men financing and constructing the lines, by Vanderbilt consolidating them and Crédit Mobilier's insiders looting the contracts, in a specific window, off subsidies and stock and steel.

Getting rich while the infrastructure is being built and keeping the returns after it becomes a regulated commodity are two entirely different achievements, and the railroads are proof that you can pull off the first and completely miss the second. When someone points at the chipmaker minting money this year as evidence that the grid owners take the whole pot, the honest answer is that the railroads had their arms merchants and their construction barons too, and they got fabulously rich, but it told you almost nothing about who would still be collecting rents in 1920.

The lesson of this chapter, and of the entire book, is in who built the lasting fortunes. They were the ones who treated cheap transport as a free input and went and built something on top of it: the steel supplier and the oil shipper and the mail-order merchant and the meatpacker and the city at the junction. They are the direct ancestors of the argument I will keep making, that the durable value in the AI build-out is going to accrue to whoever owns the data, the workflow, the customer relationship, and the reorganized business process rather than to whoever owns the compute. Sears did not own the railroad. It owned the catalog and the customer. That turned out to be the better thing to own, and it was not even a close race.

The men at Promontory in 1869 thought they were watching the railroad business reach its summit. They were half right. They were watching an infrastructure reach the point of becoming indispensable, which is a real and world-changing thing to witness. What they could not see, standing on the achievement, was that becoming indispensable was the moment the value started leaving, headed up and out toward everyone clever enough to build on what they had made.

Cheap distance did not stay a distance business; it turned into every business that distance made possible. Jevons's old paradox, that a resource getting cheap gets used in ways nobody planned for, is the reason. Cheap intelligence will do the same thing, and it will do it faster. The gold spike was real. So was the gold. It just ended up in other people's hands. The interesting question is whether the people building the intelligence grid have any better idea than the railroad men did about where their gold is going.

Chapter 4

The Monopoly That Asked to Be Regulated

> Competition and regulation are inconsistent with each other.
>
> —Theodore N. Vail, president of AT&T, in the company's annual report

On March 10, 1876, in a crowded, if not downright messy, laboratory in Boston, Alexander Graham Bell and his assistant, Thomas Augustus Watson, repeated an experiment they had tried many times before. Separated by a distance that ensured they would not hear each other otherwise, Bell was going to try to talk into a transmitter and electronically transmit his voice to Watson, who was in another room at a receiver connected to Bell's transmitter by wire. The story goes that before they tried the experiment this day, Bell spilled some acidic solution on himself and exclaimed, "Mr. Watson, come here. I want to see you." Neither man could have known they had just opened the era of electronic voice communication.

You've probably heard this story before, or perhaps you saw the movie version of the event starring Don Ameche. But what you may not know is that Bell was actually granted a patent for his telephone just three days before on March 7, 1876. He bested, by just a few hours, a rival named Elisha Gray, who filed a claim of his own but too late. A few hours made all the difference. The patent granted was US patent number 174,465. It has been called the most valuable patent ever issued.

For the next seventeen years, events unfolded exactly as you would expect when one holds an enforceable patent over a technology. The Bell Telephone Company had a legal monopoly, and it used the time to string wire, open exchanges, and establish the telephone as something more than a rich man's novelty.

The first central exchange opened in New Haven, Connecticut, in 1878, serving twenty-one subscribers, and that unglamorous little switchboard was the whole idea in miniature. Without a central exchange, every telephone would need a direct wire to every other telephone it wished to reach, and the system would strangle on its own cabling before it ever got anywhere. The exchange is what turned a collection of instruments into a network.

Then Bell's basic patents expired in the early 1890s, and the legal wall fell down. Chaos ensued. Into that opening stepped Theodore Newton Vail, who recognized what the railroad men had not.

The railroads had to be dragged into the trap. Farmers organized, legislatures acted, the Supreme Court cleared the way, and a federal commission finally sat the industry down and told it what it was allowed to charge. Nobody in the railroad business woke up one morning and decided that the smart move was to invite the government in to cap their returns. The regulation was done to them, over their objection, because they had become too essential to be left alone.

Vail saw all of that coming and walked into it on purpose.

As it happened, Vail ran the American Telephone and Telegraph Company at the two moments that mattered most. He is not a household name the way Carnegie or Rockefeller or Edison are, which is itself a small injustice, because Vail did something none of them did. He looked at an industry that was becoming indispensable and understood that indispensability was going to summon the government whether he liked it or not. Vail decided that rather than fight that logic, he would embrace it, shape it, and turn it to his advantage. He traded the dream of unlimited profit for the security of a protected, regulated monopoly. By doing so, he built the largest and most stable corporation in the world and kept it that way for the better part of a century. One biographer put it simply: Alexander Graham Bell invented the telephone, and Theodore Vail invented the telephone business.

The story of how he did it, and of what happened to all the value he sat on top of, is the second instance of the Indispensability Trap. It rhymes with the railroad experience in its bones, but the melody is different, because this time the man in charge understood the tune.

To understand what Vail built, you must understand the mess he built it out of.

The Bell patents lapsed in 1893 and 1894, in two stages. What rushed in was not orderly competition so much as a stampede. Something like six thousand independent telephone companies sprang up across the country in the years that followed. They were especially thick in the rural West and Midwest, in the small towns and farm country where Bell had never bothered to build because the density was too low and the profits too thin.

By 1902 there were more than three thousand non-Bell systems in operation. By 1907 the independents controlled very nearly half of all the telephones in the US, and at certain moments, they collectively operated more telephones than Bell did. The company that had invented the industry was, fifteen years after its patents lapsed, in danger of being a minority player in it.

The competition produced something worse than lost market share for Bell. It produced chaos for everyone. Because the Bell system and the independents refused to interconnect, a telephone was only as useful as the particular network it sat on. If your business was on the Bell exchange and your best customer was on the independent exchange across town, you could not call them.

The solution, for anyone who could afford it, was to subscribe to both, which meant two instruments on the desk, two wires running to the building, two bills at the end of the month. Businesses did it because they had to. Ordinary people mostly chose one and lived with the gaps. This was the "dual service" era, and it was a daily, tangible demonstration that a communications network divided against itself is a poor thing. A telephone system's whole value comes from its reach, from the number of people it can connect to. Split that reach in half and you cut the value of every telephone on both networks.

Bell fought the independents with every weapon it had. It filed more than six hundred patent-infringement suits. It refused to let independents connect to its long-distance lines, which were protected by a further thicket of patents that the independents could not get around, so that no independent could offer a customer the ability to call another city. Western Electric, Bell's manufacturing arm, refused to sell equipment to the upstarts.

These were the tactics of a company defending a monopoly it felt slipping away, and they were not winning. What Bell needed was not a better weapon. It needed a different idea about what the telephone business was for. That idea arrived in the person of a sixty-two-year-old man who had already retired from the company once.

The Man from the Railway Mail

Theodore Vail did not come from the telephone world, and he did not come from the world of finance. He came, appropriately enough, from the railroads.

More precisely, he came from the Railway Mail Service, the branch of the Post Office responsible for sorting and moving the mail across the country on trains. Vail joined it as a young clerk in 1869, working the line between Omaha and Ogden, which is to say he began his career riding the freshly completed transcontinental railroad, sorting letters as the country he would later wire together unspooled past the window. He turned out to have a genius for it.

During the great snow blockade of 1870, when the mail piled up and the system seized, Vail found ways to keep it moving, and the performance got him noticed. He invented what was called Fast Mail, the first train dedicated entirely to mail, running between New York and Chicago. He devised new schemes for how mail should be sorted in transit so it arrived days sooner. By February of 1876, the same year Bell patented the telephone, Vail had risen to General Superintendent of the entire Railway Mail Service, the youngest officer in the service, running the nervous system that moved information across a continent by rail.

This is the bridge from the last chapter to this one. Vail's first great feat was organizing the flow of information over the railroad network. His second and greater feat would be organizing the flow of information over the telephone network. The same mind that systematized the mail trains would systematize the wire. He understood, earlier and more deeply than almost anyone, that these networks were not collections of hardware but living systems whose value lay in their coordination, their standardization, their universality. A letter is worthless if it cannot reach its destination. So is a phone call.

He came to the telephone almost by accident. Gardiner Hubbard, Bell's father-in-law and one of the financial backers of the young company, was a congressman who sat on the postal committee. He met Vail during an inspection tour of the mail system. Hubbard was the kind of promoter who carried a pair of telephones with him to demonstrate at every opportunity, and when he pulled them out for Vail, Vail grasped at once what the potential was. Hubbard, needing a general manager for the fledgling business, hired him in 1878. When Vail started, there were about ten thousand Bell telephones in the entire country.

During this first tenure as president of the Bell companies, from 1878 to 1887, Vail did the foundational work. He built out the long-distance service, connecting Boston to Providence and then reaching further. He defended the Bell patents against Western Union, the telegraph giant that had tried to crush the telephone in its cradle. And he made the decision that would matter most in the long run, though nobody could have known it then: He set up Western Electric as the Bell system's manufacturing arm. The company that ran the network also built the equipment the network ran on. He was constructing a vertically integrated whole, the network and the factory under one roof, at a time when the multidivisional corporation barely existed as a concept.

And then he left. In 1887 Vail resigned, passed over for the presidency of the parent company and worn down by fights with the Boston financiers who backed the firm. Those investors wanted dividends. They looked at a company earning a million dollars and wanted six hundred thousand of it paid out to shareholders now. Vail wanted to plow it back into building the network for a future the financiers could not see. It is a tension that happens again and again, even today. The one between the people who own a thing and the people who are trying to build it, and in 1887 the owners won, and Vail walked away.

He spent the next twenty years as a kind of gentleman capitalist of infrastructure. He went to Argentina and built a hydroelectric plant in Córdoba and an electric street railway in Buenos Aires. He invested in mining. He turned a farm in Vermont into a model agricultural estate. There is a quiet irony in the fact that Vail spent his exile from the telephone business building the exact archetypes this book is about, electric utilities and railways, the indispensable grids of their day. Then, in the space of two years, his wife died and his only son died of typhoid, and a widowed, childless Vail, past sixty, was available when the telephone business came looking for him again.

It came looking because it was in trouble, and because J. P. Morgan had decided he was the man to fix it. The same J. P. Morgan who came to the "rescue" of the bankrupt railroads.

The Bargain

By 1907 AT&T was financially battered and strategically adrift. A bond sale the year before, led by a Morgan syndicate, had flopped so badly that only about a seventh of the offering sold, a market verdict on a company that had spent heavily, innovated little, and watched independents take half its market. The Morgan interests moved in, took control, and installed the one man they believed could turn it around. On the first of May 1907, Theodore Vail became president of the Bell companies, now called AT&T, for the second time, and this time he had both the authority and the backing to do things his way.

What he did first was reverse the war. Under Morgan's orders he cut costs hard, firing twelve thousand employees to stabilize the finances. But the strategic turn was the thing that mattered. Vail stopped trying to crush the independents and started trying to absorb them. When the financial panic of October 1907 pushed many small telephone companies to the edge of collapse, Vail was there not to bury them but to buy them, folding them into the Bell system one after another. Where he could not buy, he offered to connect. An independent could link to Bell's long-distance network for a fee, and suddenly its customers could call the whole country. The independents that took the deal became, in effect, extensions of the Bell system. The independents that refused found themselves marooned on islands while Bell offered the mainland.

Underneath the tactics was a philosophy that Vail spelled out in a series of annual reports that are unlike anything else in the history of American business writing. They are essays, really, personal and discursive and occasionally pontifical, in which the head of the largest telephone company in the world thought out loud about the proper relationship between a corporation and the public. In the 1907 report, he introduced a section he titled "Public Relations," by which he did not mean advertising but the entire question of how a company this important ought to conduct itself toward the society that depended on it. And he introduced the slogan that would become the company's creed for seventy years: One Policy, One System, Universal Service.

The genius of Vail's position was that he made an argument the government wanted to hear, and he meant it. He argued that the telephone was a natural monopoly, that competition in telephone service produced not efficiency but waste and chaos, the two-instruments-on-the-desk absurdity of the dual-service era. Duplicating networks was foolish; the public was best served by one unified, interconnected system.

But, he went on, a monopoly of something so essential could not be allowed to behave like an ordinary business, squeezing every dollar it could from a captive public. A monopoly on a public necessity carried obligations. It had to serve everyone, it had to be reliable, and it had to accept public regulation of its rates and its conduct in exchange for its protected position. Vail said, in as many words, that competition and regulation were inconsistent with each other, and having framed the choice that way, he chose regulation. He even argued, remarkably for the head of a for-profit enterprise, that maximum private profit was not the primary purpose of the company. Profit was necessary to keep the enterprise healthy, but it was one element in an equation, not the whole of it.

You can read this cynically, and plenty have. A monopolist who embraces regulation is a monopolist who has found a way to make his monopoly permanent and respectable, to get the government to bless the very market dominance it might otherwise attack. There is truth in that reading. But it misses what makes Vail important for our purposes, which is that he understood the Indispensability Trap and made terms with it before he was forced to. He saw that the telephone was becoming too essential to escape the government's attention, and rather than wait to be dragged before a commission like the railroads, he walked in, sat down, and negotiated.

The negotiation came to a head in 1913. AT&T's acquisitions, and Vail's move to buy a controlling interest in Western Union and reunite telephone and telegraph under one roof, had grown alarming enough that the federal government readied an antitrust suit. Vail did not fight it. Through a vice president named Nathan Kingsbury, AT&T made a settlement with the Justice Department out of court that has been known ever since as the Kingsbury Commitment. Its terms: AT&T would sell off its controlling stake in Western Union. It would allow independent telephone companies to interconnect with its long-distance network. And it would stop acquiring competing independents without government approval.

On its face Kingsbury looks like a defeat, concessions extracted from a monopolist by the threat of prosecution. In the deeper logic, it was the opposite. By accepting these limits, AT&T defused the antitrust threat, legitimized its dominant position, and secured its future as a regulated monopoly. Historians of the industry routinely describe the Kingsbury Commitment as the beginning of AT&T's monopoly, not the end of it, because it was the moment the company traded the risks of unchecked expansion for the security of a protected, sanctioned dominance. Vail gave up the freedom to charge whatever he wanted and buy whomever he pleased, and in exchange he got something a businessman almost never gets: a durable, defensible, government-blessed monopoly on one of the most important technologies of the century. The cap on his upside was the price of admission, and he paid it gladly.

The arrangement hardened over the following decades. The Willis Graham Act of 1921 exempted telephone mergers from antitrust, letting Bell resume its consolidation. The Communications Act of 1934 created the Federal Communications Commission and formalized AT&T's status as a regulated monopoly overseen by the federal government. For half a century after Vail's bargain, AT&T was the phone company, singular, a regulated utility whose rates were set in negotiation with commissions and whose reach extended to nearly every home and business in America. Vail had been right. The trap was coming. He simply chose to build his house inside it.

The Countryside

There was a second half to Vail's bargain, and it is the half that recurs in several chapters, so let me name it here.

When a technology becomes essential, society does two things at once, and they are really the same thing. It caps what the owner can charge, and it insists that the essential thing reach just about everyone, including the people it would never be profitable to serve. Those two moves are inseparable, because you cannot promise a monopolist unlimited profit and universal access in the same breath. The one is paid for by the other.

Vail's slogan contained the word "universal," and it is important to be precise about what he meant by it, because the meaning has shifted under the phrase over the years. When Vail said "universal service" in 1907, he did not mean what a modern reader hears, subsidized telephone service reaching every farmhouse in the nation. He meant something closer to "one interconnected system," universal in the sense of unified rather than fragmented, the cure for the dual-service chaos. Universal service, to Vail, was the opposite of the two-phones-on-the-desk world. It was the promise that the network would be whole.

But the phrase turned out to be larger than its author, and the modern meaning grew into it. The Communications Act of 1934 charged the new FCC with making telephone service available, in the law's words, so far as possible to all the people of the United States at reasonable charges. That language became the root of what we now call the universal service principle, the idea that access to the network is something like a right, and that the market, left alone, will never extend it to the rural and the poor and the remote because there is no money in it.

The way this actually worked for most of the twentieth century is one of the quiet marvels of the regulated era, and almost nobody noticed it happening. The regulated monopoly was the instrument of universal access, through a vast, deliberate system of cross-subsidies. The regulator allowed AT&T to charge more than cost on the services where it could, on long-distance calls, on business lines, on urban service, and required it to charge less than cost where society wanted the network to reach, on local residential service, on rural lines strung out to farms at a loss. The profitable parts of the system carried the unprofitable parts. A businessman in Chicago placing a long-distance call to New York was, without knowing it, helping pay to run a telephone line to a farmhouse in Nebraska. The Supreme Court, writing about this history as recently as 2025, described it plainly: The FCC pursued universal service through implicit subsidies, using its rate-setting power to lower costs for some by raising them for others.

This is the guarantee-access half of the Indispensability Trap made concrete. The cap on AT&T's returns and the extension of service to unprofitable places were not two separate policies. They were one bargain. AT&T accepted regulated, limited profit, and in exchange it was handed a protected monopoly and made the vehicle through which the whole country got wired.

When we reach electrification in the next chapter, we will watch the same bargain be struck again, this time with the government building the rural lines directly because the private utilities would not. When we reach the internet, we will watch it a third time, fought out over broadband and the digital divide. With regard to artificial intelligence, the question of whether cheap, capable intelligence will be guaranteed to everyone or left to reach only the profitable is still to be settled.

Up the Stack, Part One: The Giant's Own Laboratory

This next part of the telephone story complicates the simple version of my thesis, but it rewards the complication with something richer.

With the railroads, the value migration was clean. The operators went broke and the fortunes went elsewhere, to Carnegie and Rockefeller and Sears, to people who did not own track. AT&T did not go broke. It became the largest and most stable company on earth and stayed that way for generations. So where, in the telephone story, is the value climbing off the grid and into other hands?

Part of the answer is that AT&T itself climbed up the stack, and this is the part most people do not know or recognize. The regulated phone monopoly was not only a phone company. It owned Western Electric, which was for decades one of the largest industrial manufacturers in the United States, the company that made virtually every piece of equipment in the Bell system. And in 1925 it created, jointly with Western Electric, an institution called Bell Telephone Laboratories, which would become the most productive industrial research laboratory in the history of the world.

The list of what came out of Bell Labs is difficult to believe when you see it assembled in one place. The transistor, invented there in 1947, the device that every piece of modern electronics is built on, the thing that made the computer age physically possible. Information theory, invented essentially single-handedly by Claude Shannon in 1948, the mathematical foundation of all digital communication, the reason we can speak of "bits" at all. The laser. The solar cell. The charge-coupled device that sits inside every digital camera. Radio astronomy and the accidental discovery of the radiation left over from the Big Bang. Communication satellites. The Unix operating system and the C programming language, invented at the Labs around 1970, which together became the substrate of modern computing and made the internet practical. The people who did this work collected eleven Nobel Prizes and five Turing Awards among them. There has never been anything else quite like it.

The reason Bell Labs could exist, the reason it could fund brilliant people to chase fundamental questions with only the loosest connection to selling telephone service, was precisely that AT&T was a regulated monopoly with guaranteed, stable, capped returns. A company fighting for its life in a competitive market cannot afford to employ a mathematician to work out the theoretical limits of communication for its own sake. A protected utility with a predictable income and a mandate to think long term can. The cap on AT&T's profit did not starve its capacity to invent. It funded it. The same regulated stability that limited how much money AT&T could make gave it the patient capital to build the greatest engine of invention the century produced. That is a genuinely surprising turn in the logic of the trap, and it is worth holding onto: The constraint that limited the winner also bankrolled the future.

You could see the machine warming up early, in the achievement that first showed what AT&T's research muscle could do. On January 25, 1915, the company opened the first transcontinental telephone line and staged the opening as a piece of theater. Alexander Graham Bell, now sixty-seven, sat at one end in New York. Thomas Watson, now sixty, sat at the other in San Francisco. Bell spoke into the telephone the same words he had spoken to Watson thirty-nine years earlier in the first phone call ever made, "Mr. Watson, come here. I want to see you," except that this time the voice traveled three thousand miles instead of the length of a room.

The line was made possible by two inventions that let a signal cross a continent without fading into nothing, a loading coil and a vacuum-tube repeater, the repeater built from a device AT&T had bought and its engineers had refined. The last pole of that line, as it happens, went up at Wendover, on the Utah-Nevada border, in the same desert where the previous chapter began with a golden spike. The rails were joined in Utah in 1869; the wire was joined in Utah in 1915. The grids of America keep finishing themselves in the same empty country.

Up the Stack, Part Two: The Fortune They Were Forbidden to Keep

But if AT&T went up the stack through Bell Labs, the deeper truth of the telephone story is that the company was not allowed to keep most of what it found up there. And this is where the value migration reasserts itself, more dramatically than in any other chapter.

In 1949 the government brought another antitrust suit against AT&T, and in 1956 it was settled with a consent decree that, in hindsight, may be the most consequential document in the history of American technology. Under it the Eisenhower administration let AT&T keep Western Electric, which the company badly wanted. In exchange AT&T agreed to two things. It agreed to stay out of any business other than regulated telephone service, which meant, specifically and fatefully, that it would not enter the computer industry. And it agreed to license the patents it already held, including the transistor, royalty-free, and to license its future patents at reasonable royalties.

Consider what that meant. Bell Labs had invented the transistor, the foundational component of the entire electronic age. And the company that owned it was now legally required to give it away for nothing and legally forbidden from entering the industry that the transistor would create.

The result was one of the great unintentional acts of technology transfer in history. The transistor, free for the taking, seeded a new industry that grew up mostly in California, in the semiconductor firms that became Fairchild and then Intel and then the whole of what we call Silicon Valley. AT&T fathered the computer age but was barred by law from profiting from it. The value it had created did not stay in the laboratory that created it. It migrated, wholesale, into the hands of everyone else, into an industry the inventor was not permitted to join.

This is my thesis in its purest and most painful form. The owner of the grid can invent the future and still be unable to keep it. Sometimes the value migrates because the market pulls it up the stack to the builders on top. And sometimes, as here, it migrates because the state reaches in and pries it loose, deciding that a thing this important cannot be allowed to stay in one company's hands. Either way, it goes, and it went.

Think about the counterfactual for a moment, because it is staggering. A version of AT&T that had been free to build on its own inventions, to enter the computer business it had made possible, to charge for the transistor it had discovered, might have become the largest enterprise the world has ever seen. Instead, operating with the government's sword permanently over its head, it watched the industries it had fathered grow up in other people's houses.

And there is a second, quieter migration in the telephone story, one that looks most like the mail-order merchants of the last chapter. All across the twentieth-century economy, entire businesses grew up on top of cheap, ubiquitous telephone service, owning no wire themselves. The call center. The toll-free 800 number, which turned the telephone into a channel for commerce and let a company in one state take orders from customers in all of them. Catalog retail found a second life on the phone. Telemarketing was born, for better and mostly for worse. Customer service became a department and then an industry. The fax machine reshaped how businesses moved paper for a generation. None of these enterprises owned the network. They simply assumed it, the way Sears assumed the railroad and built on top of the assumption.

The largest of these build-on-top stories was the one nobody planned. When the internet arrived at the ordinary home in the 1990s, it did not arrive over some new infrastructure built for the purpose. It arrived, screeching and hissing through a modem, over the telephone line. The dial-up internet, and then the DSL broadband that followed, ran on the copper that the phone company and its successors had already strung to nearly every building in America, the local loop, the last mile, the physical legacy of a century of universal service. The entire early internet economy was, quite literally, an application running on top of the telephone network. And the value that migrated onto that network, the value of everything the internet became, would eventually dwarf the value of the network itself by orders of magnitude. That is the next chapter's foreshadowing, and the one after that, and, if this book is right, our own.

The Sword Falls

The Indispensability Trap has one more mode of enforcement beyond rate regulation, and the telephone story is where we see it used. When a company becomes so completely dominant, so total in its control of something essential, that even a regulated monopoly seems too much power in one place. The state does not merely cap it; it breaks it apart.

The fear that drove this was the old one, the same fear the railroads provoked. It was not a fear of the telephone itself but a fear of the concentration of power, of a single private entity holding a chokehold on the nation's communication.

By the 1970s AT&T reached 90 percent of American households and controlled essentially the entire long-distance network, the manufacturing of the equipment, and the research besides. In 1974 the Justice Department filed the antitrust suit that would end the Bell system. It took years. The trial before Judge Harold Greene began in 1981. On the eighth of January 1982, a day some in the company called Fateful Friday, AT&T's chair and the government's antitrust chief announced a settlement.

The terms were a kind of amputation. AT&T would divest its local operating companies, the parts of the business that actually connected to people's homes, which amounted to roughly two-thirds of the company's assets. Those local companies were reorganized into seven regional holding companies that the press immediately nicknamed the Baby Bells. AT&T itself would keep the long-distance network, Western Electric, and Bell Labs, and, freed at last from the 1956 decree, would be allowed to enter the computer business it had been shut out of for a quarter century. On the first of January 1984, the Bell system, the thing Vail had spent his life building into one unified whole, ceased to exist, carved into eight pieces along the lines a federal judge had drawn.

Look at what the breakup did to Vail's bargain. The cross-subsidies that had funded universal service ran through the old integrated monopoly, long-distance revenue quietly holding down local rates. Break the company in two, separate the long-distance business from the local one, and the subsidy has nowhere to flow. Local telephone rates rose after 1984. The people the old system had protected, the residential and rural customers whose service had been carried by the profitable parts of the network, were among the ones who paid for the new competition. The bargain, it turned out, had been load-bearing.

Humpty Dumpty Reassembles Himself

What happened next surely put a smile on Vail's face, wherever he was. The pieces did not stay broken.

AT&T's grand plan for its liberation went wrong almost immediately. Freed to enter the computer business, it tried, and failed, unable to make itself into a computer company by force of will. Western Electric, the mighty manufacturer, turned out to be unprofitable once it no longer had a guaranteed captive customer in a Bell system that was required to buy from it. By the mid-1990s AT&T voluntarily did what the government had originally wanted, and it had resisted. It spun off its computer operations and its manufacturing arm entirely. Western Electric and Bell Labs became a new company called Lucent Technologies, which a decade later merged into Alcatel and was eventually swallowed by Nokia, so that Bell Labs, the greatest laboratory of the American century, is today a research division of a Finnish telecommunications firm.

And the Baby Bells, the seven regional companies created to ensure competition, spent the next two decades doing the one thing the breakup was supposed to prevent. They merged. They bought each other. They reassembled. Southwestern Bell, one of the seven, grew and acquired and eventually, in 2005, bought its own former parent, the long-distance AT&T, and took the storied name for itself. Bell Atlantic combined with others to become Verizon. When the dust settled, the seven Baby Bells and the old AT&T had recombined into essentially two national giants, AT&T and Verizon, each of them once again offering local service and long-distance, and now, wireless and broadband and television, the very combination of everything-under-one-roof that the 1984 breakup had been engineered to forbid. The structural separation that a federal judge had so carefully designed lasted barely a generation. As recently as early 2026, AT&T completed a $5.75 billion purchase of Lumen's consumer fiber business, assets that traced back through Qwest to US West, one of the seven Baby Bells.

You can draw two lessons from the reassembly and both are important to my thesis. The first is that the economics of network infrastructure pull relentlessly toward concentration, that the natural monopoly logic Vail identified in 1907 did not go away because a court wished it away, and that when you break up an essential network, the pieces tend to find their way back together because the underlying forces never changed. The second is subtler, and it is the one that matters most for where this book is going. After all of it, the reassembly, the return to dominance, AT&T and Verizon are not where the great fortunes of the internet age were made. They own the wires that the internet still runs on. They are indispensable, regulated, enormous, and permanently capped. The trillions in value that rode across their networks over the past thirty years accrued somewhere else, to the companies that built on top of the connection rather than owning it. The grid put itself back together and remained, precisely, the grid.

It Wasn't Just America

Before we close this lesson, it is worth stepping outside the US for a moment, because a skeptic could fairly object that everything discussed so far is a story about the peculiarities of American antitrust politics rather than a law of technology. The objection dissolves the instant you look abroad, because nearly every other developed nation ran the same play by the opposite means and arrived at exactly the same place.

Most of the world did not bother with the American compromise of a regulated private monopoly. It simply made the telephone company an arm of the state. Across much of Europe, including France, the pattern was the PTT (the Post, Telegraph, and Telephone ministry), a government department that ran the mail, the telegraph, and the telephone together as a single public monopoly. Britain ran its telephones through the General Post Office for the better part of a century. Germany had the Deutsche Bundespost and Japan had NTT. In each case the recognition was the same one Vail had, that the telephone was too essential to leave to an unregulated market, but the conclusion was more direct. Where America capped the monopoly with regulation, these countries skipped the intermediary and owned the monopoly outright. State ownership was not an alternative to the Indispensability Trap; it was the trap in its most naked form, the technology grown so essential that the government simply took it.

And Britain shows the trap's dark side, the risk the American arrangement partly escaped. A government department has no competitor and no strong reason to hurry. The General Post Office throttled the technology for decades, leaving Britain notoriously under-telephoned compared to America. A state monopoly that raised the money it needed by raising rates rather than by getting better. The cap that funded Bell Labs on one side of the Atlantic produced complacency on the other. The trap, depending on how it is run, can build a cathedral of research or produce a sleepy bureaucracy, and that is a lesson worth carrying into the argument about who should own the intelligence grid.

Then came a telling coincidence. In the same decade that the US broke up AT&T, the state monopolies were privatized and broken open too, and Britain led the way in the very same year. In November 1984, ten months after the Bell system was dissolved, Margaret Thatcher's government sold British Telecom to the public for £3.6 billion, the largest stock flotation in history to that point and the template that would launch the privatization of public infrastructure across the world. France and Germany followed over the next decade. Two systems, built on opposite ideologies, one capping a private monopoly and the other owning a public one, arrived at the same destination by the same route at the same time: monopoly, then a cap, then liberalization, and then the value migrating to whatever got built on top.

When two philosophies that agree on nothing else produce the same history, you are no longer looking at politics. You are looking at something structural.

What Vail Understood

Set the whole arc down and look at it, and the six questions I put to every case in this book answer themselves.

How did the telephone cross from novelty to necessity? Through the exchange and the network, until a nation that had lived without telephones could no longer imagine doing so, and dual-service fragmentation became intolerable precisely because the thing had become essential.

Who paid to build it, and did they keep the wealth? Vail and Morgan built it, and AT&T kept an enormous amount of wealth, more than any operator in this book, but kept it as a regulated utility with its upside permanently capped, and the truly uncapped fortunes climbed off the network onto the layer above.

Did it commoditize and get regulated? It was regulated more thoroughly and for longer than almost any enterprise in American history, and this time by the owner's invitation.

Where did the durable value go? Up the stack, to Silicon Valley by way of a transistor its inventor was forbidden to keep, and to the entire internet economy that rode across the wires.

What did people fear? The concentration of power, as with the railroads, answered with antitrust. And how much of this holds for artificial intelligence? A great deal, so let's be specific about what that is.

The first transfer is the one that gives this chapter its title. The trap can be chosen. Vail is proof that the smartest response to becoming indispensable is not to fight the regulation that indispensability summons but to shape it, to trade unlimited upside for a durable, protected, sanctioned position. Somewhere in the AI build-out there is a company that will do what Vail did, that will read the trap coming and walk into it on purpose, embracing regulation as strategy while its rivals get dragged. Watch for it. It may already be maneuvering.

The second transfer is the strangest and the most hopeful. The cap can fund the future. AT&T's regulated, limited, guaranteed returns paid for Bell Labs and the transistor and information theory and Unix, the patient, long-term research that a company in a knife fight can never afford. If artificial intelligence settles into a regulated, capped, utility-like business at the infrastructure layer, that is not only a story of diminished returns. It may also be the condition under which someone builds the Bell Labs of intelligence, the great, patient laboratory that a frantic competitive market would never sustain. The constraint and the cathedral came together once. They could again.

The third transfer is the one this whole book turns on. Value migrates even from inside the winner. AT&T fathered the computer age and was made to give it away. The internet grew up on wires AT&T owned and could not monetize beyond the toll for the connection. The company that built the grid watched the century get built on top of it. If the largest, most dominant, most vertically integrated infrastructure monopoly in the history of the world could not hold onto the value it created, could not keep the industries its own laboratory invented, then the hyperscalers pouring their hundreds of billions into the intelligence grid should look very carefully at the ninety years between Vail's bargain and Verizon and ask themselves a hard question. Are they building something magnificent and essential? If so, they will likely be regulated or broken up or simply out-migrated by the builders on top, and possibly all three. The one thing the telephone century suggests they will not get to do is keep it all.

There is one company today with the assets in AI to attempt what AT&T attempted, to own the grid and the laboratory and the layer above it all at once, to be the phone company and Western Electric and Bell Labs of the age of intelligence. It has the network, the research arm that already wins the science prizes, the models, and the distribution to a few billion people. Whether it can hold that position, or whether the same forces that pried the transistor out of AT&T's hands will pry the future out of its own, is one of the real questions of the coming decade. The telephone century does not tell us the answer. It only tells us, with a hundred years of evidence, which way to bet. If you don't know which company that is, I will reveal it in the second half of the book when we talk about AI.

Vail understood all of this, or enough of it. He understood that the telephone business was not the same as the telephone, that the wire was not the wealth, that becoming essential was a kind of golden cage. He built the most successful cage in the history of business and lived comfortably inside it for the rest of his life. What he could not do, what no one in his position has ever quite managed to do, was keep the fortune from climbing over the walls.

Chapter 5

The Man Who Asked to Be Regulated

> The obligations of monopoly must be accepted.
> —Samuel Insull, 1910

On the morning of August 6, 1890, in a room at Auburn Prison in upstate New York, a carpenter's assistant and convicted murderer named William Kemmler was strapped into a wooden chair that had been built, in secret and at some expense, to kill him with electricity. The chair was the first of its kind in the world. The men who had championed it promised that death by electric current would be instant and humane, a scientific improvement over the crude violence of the noose. The current was switched on for seventeen seconds, but Kemmler did not die. Witnesses realized he was still breathing, and the generator had to be run back up to speed while he convulsed in the chair. The second jolt lasted long enough that the room filled with the smell of scorched flesh. By some accounts, smoke rose from his body. It took about eight minutes. George Westinghouse, who had followed the whole grim affair from a distance, offered the epitaph: "They could have done better with an axe."

What almost no one in that room understood was that they were not merely witnessing an execution. They were witnessing a move in a business war. The man who had quietly arranged for that chair to run on alternating current, and had secretly paid to make sure it did, was Thomas Edison.

This is the strangest and darkest episode in the story of how America got wired. I include it because electrification is the technology that the artificial intelligence build-out rhymes with most. The War of the Currents is the part of the story that speaks most directly to our own anxieties.

The fear within each build-out so far has been a fear of concentrated economic power. This one is about a fear of the technology itself. A fear that the thing being built into the walls of every home in America might simply kill the people living there. That fear was not irrational. Electricity, badly handled, does kill. And the reason the story matters for us is that the fear turned out to be both completely real and completely survivable, which is the exact question hanging over artificial intelligence today.

But the War of the Currents, for all its theater, is not the heart of this chapter. Edison and Tesla and Westinghouse built the stage. The man who walks out onto it and turns a miraculous invention into an industry, and in doing so becomes the truest example in this entire book of the pattern we are tracing, is a bespectacled, five-foot-three English immigrant. He came to America as Edison's secretary and died, decades later, on the stairs of the Paris subway. His name was Samuel Insull, and to understand him you first have to understand the war he watched the other men fight.

The War of the Currents

By the late 1880s the electric light had been proven, and the question was no longer whether America would be electrified but how. Two systems competed to become the standard, and the difference between them was not a matter of taste. It was physics, and the physics had enormous commercial consequences.

Edison had built his empire on direct current, in which electricity flows steadily in one direction at the same low voltage the customer's lamps used. Direct current (DC) worked, and Edison had staked everything on it, but it had a crippling limitation. It could not be transmitted very far. Low voltage current fades quickly over distance, so a direct current system needed a generating station every mile or so and thick, expensive copper wires to carry the load. To electrify a city with DC, you needed to blanket it with power plants. To electrify the countryside? You essentially couldn't.

Alternating current (AC) solved exactly this problem. In an AC system the current rapidly reverses direction. Crucially, its voltage can be stepped up very high for transmission over thin, cheap wires across long distances, then stepped back down to a safe level for use. This is the technology, developed by Nikola Tesla and commercialized by George Westinghouse, that made it possible to build one large generating station and serve a whole region from it. It is, with refinements, the system the entire world still runs on. On the engineering merits, the contest was close to settled before it began. AC was simply the better way to move power.

Edison knew it, and rather than concede he went to war. The war he chose to fight was a campaign of fear. If AC was going to beat DC on economics, Edison would beat AC on terror. He set out to convince the American public that alternating current was a lethal danger waiting in the walls of their homes.

The campaign was as grim as anything in the history of American business. Edison hired an engineer named Harold Brown and paid him in secret to stage public demonstrations of AC's deadliness. Brown and his associates electrocuted animals before crowds and reporters, using alternating current to do it. Neighborhood boys were paid a quarter apiece to round up stray dogs, which were then killed with AC in front of audiences. The demonstrations escalated to cats, then to horses and cattle. Edison's people coined a word for death by electrocution, proposing that the public call it being "Westinghoused," after his rival. Brown published a booklet, very likely paid for by Edison's company, purporting to prove with all these animal experiments that AC was far deadlier than DC. They then saw that it was mailed to officials and newspapers across the country. Brown was later accused of rigging the tests, quietly using low-current DC and high-current AC to manufacture the result he wanted.

The campaign reached its infamous peak with the electric chair. When New York State went looking for a more humane alternative to hanging, Edison, who had previously called himself an opponent of capital punishment, saw an opportunity too dark to pass up. If the state could be persuaded to execute its condemned prisoners with alternating current, then AC would be forever associated in the public mind with death. No family would want it running through their home. Edison threw his prestige behind electrocution as the humane method, and behind Brown's effort to build the chair, and made sure it would run on AC. Westinghouse, understanding exactly what was being done to his technology, refused to sell his AC generators for the purpose. Brown obtained them anyway, through back channels. Thus William Kemmler died badly in Auburn Prison. Westinghouse, watching his life's commercial work turned into an instrument of botched death, delivered his line about the axe.

To be fair to the historical record here, the popular version of this story has grown some barnacles. The animal experiments did not spring purely from Edison's malice; they took place against a genuine, and genuinely grisly, contemporary effort to find humane methods of killing unwanted animals and executing criminals, prompted in part by inquiries from humane societies and a state commission. The single most famous image associated with this whole affair, the electrocution of Topsy the circus elephant at Coney Island, does not actually belong to it. Topsy was killed in 1903, years after the current war was over and Edison had already lost, and she was put down for having killed her handlers, not as a stunt against Westinghouse. The story that Edison personally electrocuted an elephant to smear alternating current is one of those urban myths everyone knows that is not so. The real story is bad enough without it.

Here is how the war ended. None of it worked. The propaganda, the dogs, the chair, the ugliness, none of it changed the physics or the economics. In 1893 Westinghouse won the contract to illuminate the World's Columbian Exposition in Chicago and lit the fair with alternating current before twenty-seven million visitors. That same year he won the far more important prize, the contract to harness Niagara Falls, and when the Niagara plant began sending alternating current to the city of Buffalo in 1896, twenty-six miles away, the argument was over. AC could do the one thing DC never could, move enormous power across real distance, and that capability was worth more than all the fear Edison could manufacture. Alternating current won, and the modern electrical grid was born from that victory.

One point carries forward, past all the spectacle, into the rest of this book. The fear was real. Alternating current genuinely can kill you. The danger Edison exploited was not invented; it was exaggerated and weaponized but not fabricated. People did die in the early decades of electrification, from shocks, fires, and accidents, as a genuinely dangerous new force was threaded into ordinary buildings. Society did not respond by rejecting electricity, nor by pretending the danger away. It responded by making the technology safe enough, through insulation, grounding, standards, inspection, and regulation, that the benefits overwhelmingly outweighed the risks.

The safety fear about electrification was true, and it was survivable, and the surviving was done not by the market alone but by the market plus a great deal of rulemaking. Remember this, because when we reach the argument about whether artificial intelligence is dangerous, and what if anything to do about it, the War of the Currents is the closest thing history offers to a precedent, and it cuts in more directions than either the boosters or the doomsayers would like.

The Secretary

For now, the stage is set, and the protagonist can enter.

In 1881 a twenty-one-year-old Englishman stepped off a boat in New York, having crossed the Atlantic to take a job he had maneuvered for with almost frightening determination. He was small, barely over five feet, and slight, and he wore spectacles and dressed with a formality that made him look like a caricature of Victorian propriety. He spoke with a thick London accent that his new employer, a man with a flat Midwestern one, could barely understand. His name was Samuel Insull, and he had come to America to be the private secretary of Thomas Edison.

Insull's whole life to that point had been a study in relentless self-improvement from modest beginnings. The son of a lay preacher and temperance campaigner, he had gone to work at fourteen as an office boy in an auction house for a little over a dollar a week and taught himself shorthand and bookkeeping at night. He read everything he could about Edison, who was already a transatlantic celebrity, and when he learned that Edison's London representative needed a secretary, he got the job and turned it into a listening post. He learned the electrical business from the correspondence that crossed his desk. And when the chance came to go to America and work for the great man himself, Insull took it over an offer of a banking career with the firm of Drexel, Morgan, because he did not want to be a banker. He wanted to work for Edison.

Edison expected a secretary and discovered, quickly, that he had hired something far more valuable: a superb businessman. Insull had a genius Edison entirely lacked, for organization, for finance, for the unglamorous machinery of turning inventions into functioning companies. Edison was chaotic, brilliant, careless with money and detail. Insull was precise, tireless, and financially ruthless. Within a brief time, the small Englishman had the complete trust of his portly employer and given effective control of Edison's business affairs, with power of attorney and the authority to run the enterprises Edison had scattered around him. There is a line about the pairing that captures the whole thing:

Edison was a genius, but a bright young Englishman named Samuel Insull made him rich.

Insull was central to the great early demonstration of what electricity could be, the Pearl Street Station in lower Manhattan, which switched on in September of 1882 and lit a square mile of the financial district from a single central plant. He helped raise the money, wrangle the city officials, some of it done with bribes, and bury the hundred thousand feet of wire the system required. Pearl Street was the proof of concept for the whole industry, the first central power station serving a district of paying customers, and Insull learned its central lesson in his bones. The economics of a power station depend entirely on keeping it busy. A plant that runs flat out is a gold mine; a plant that sits idle is a catastrophe, because the enormous cost of building it must be paid for, whether it is generating revenue or not. That insight, that the whole game is filling the plant, would later make Insull the most important figure in the American electrical industry.

But not yet, and not in New York. In 1892 the financier J. P. Morgan (here he is again) engineered the merger that created General Electric, folding Edison's manufacturing operations into a new giant. In the reshuffle Insull was offered a senior position but was passed over for the top job. He was not a man who enjoyed being second. So when the directors of a struggling utility out west, the Chicago Edison Company, asked Insull to recommend someone to run it, he gave them a recommendation that tells you everything about him. He recommended himself. He took a pay cut to do it and signed a contract designed to make it hard for him to come crawling back to New York, and in 1892 he went to Chicago to build, out of one small power company in a city served by a dozen competitors, the largest electrical empire in the world.

Massing Production

What Insull did in Chicago over the next three decades is the part of this story that matters most for understanding our own moment, and it is the part almost nobody remembers, because it is not about invention. Edison and Tesla did the inventing. Insull did something that turned out to be worth more. He figured out how to make electricity cheap.

When Insull arrived, electricity was a luxury, a marvel for the wealthy and for businesses, priced high and used sparingly. He set out, with complete deliberation, to drive the price down and the volume up, and he understood that these were the same project. His model, which he called "massing production," was rightly compared by a contemporary to what John D. Rockefeller had done with oil. The logic was a flywheel. Lower the price of electricity and more people use it. More usage justifies building bigger, more efficient generating stations. Bigger stations produce power at lower cost per unit. Lower cost lets you cut the price again, which brings in still more customers, which justifies still bigger plants. Round and round, each turn making electricity cheaper and more universal, and each turn making Insull's system larger and more dominant. He was deliberately commoditizing his own product, driving its price toward the floor, because he understood that the floor was where the volume was, and the volume was where the empire was.

The engine of the flywheel was an insight about idle capacity that Insull pursued more rigorously than anyone before him. A power plant built to handle the evening peak, when every household turns on its lights at once, sits half-asleep the rest of the day, and all that idle iron still has to be paid for. The way to make the plant cheap per unit was to keep it working around the clock, which meant finding customers whose demand came at different times of day and different seasons of the year.

So Insull chased variety on purpose. He courted daytime industrial users whose motors ran while the lights were off. Above all he courted electric traction, the streetcars and interurban railways whose enormous, steady daytime pull was the perfect complement to household lighting, and which he increasingly came to own outright. He was assembling a portfolio of customers whose peaks and valleys canceled each other out, so that his plants ran closer and closer to full all the time. He pioneered new ways of pricing that separated the charge for a customer's peak demand from the charge for their total usage, a sophistication decades ahead of its time. He ran relentless campaigns to manufacture household demand, urging Chicagoans, in one memorable slogan, to "give something electrical for Christmas."

The results were staggering, and they are the clearest illustration in this book of a principle we have already met and will meet again. As Insull made electricity cheaper, people did not use less of it. They used vastly more. The cheaper it got, the more uses people found for it, and the more they consumed, which is the pattern the economist William Stanley Jevons identified with coal decades earlier and which recurs with every foundational resource.

Cheap electricity did not satisfy some fixed demand and stop. It created an appetite for electricity that had never existed, an entire way of life built on the assumption of abundant power, and that appetite grew without limit. Commonwealth Edison, the company Insull assembled in 1907 by merging his Chicago holdings, saw its output multiply many times over across his reign. By 1920 it was burning more than two million tons of coal a year and serving half a million customers. At the peak of his empire, in the late 1920s, you could buy electricity generated by Samuel Insull's companies in some five thousand communities across thirty-two states. He was, by capital invested and power distributed, the largest producer of electricity in the world.

Notice what has happened here. It is the thing the whole book turns on. Insull did not get rich by keeping electricity scarce and expensive, the way you might expect the owner of an essential resource to behave. He got rich by making it cheap and abundant. The commoditization of electricity was not something that happened to Insull against his will. It was his strategy. And that strategy is precisely the one now driving the price of machine intelligence toward the floor. The people who tell you that cheap intelligence spells doom for the companies building the AI grid have the history backward. For Insull, cheap was the whole plan. The question was never how to keep the product expensive. It was how to own the largest possible share of an enormous, cheap, indispensable flow.

The Bargain

Now we arrive at the move that makes Insull a central figure of this book, the thing he did that connects him directly to Theodore Vail in the previous chapter and to the argument about artificial intelligence in the chapters ahead.

Samuel Insull, a monopolist who had crushed or bought nearly every competitor in his territory, went to the government and asked to be regulated.

This was not the behavior anyone expected of a man in his position, and it baffled other businessmen. The conventional monopolist fears regulation above all things and fights it with every resource. Insull invited it. In speeches and writings, most pointedly in a 1910 address whose title captured a philosophy he had first argued in 1898, "The Obligations of Monopoly Must Be Accepted," he argued that electricity was a natural monopoly, that it made no sense to have competing companies stringing duplicate wires down the same streets, and that the efficient outcome was a single provider serving each territory. But he went on, saying that a monopoly on something so essential could not simply be left to charge whatever it liked. It had to accept public oversight. It had to submit to a state commission that would regulate its rates and guarantee its service, and in exchange, it would receive a protected exclusive territory and the assurance of a fair, which is to say a limited, return on its investment.

Why would a monopolist want this? Because Insull was playing a longer game than his rivals could see. He understood that a monopoly on an indispensable service was politically intolerable if it was unregulated, that the public would eventually rise up against it. The alternative to regulation was not unfettered profit but municipal takeover, with cities seizing the power companies and running them as public utilities. Regulation was the deal that made his monopoly permanent and safe. A state commission setting his rates was a small price to pay for a legally protected territory, a guaranteed return, and immunity from both competition and confiscation. He traded the dream of unlimited profit for the security of a sanctioned, regulated, permanent monopoly, and he did it on purpose, years before anyone forced his hand.

This is the exact move Theodore Vail made with the telephone, and the two men made it independently, in different industries, in the same era. That is what turns this from an anecdote into a law. When one man reads the trap of indispensability and chooses to walk into it, you might call it a quirk of temperament. When two of the most important industrialists of the age, in two different essential industries, look at the same situation and reach the same counterintuitive conclusion, that the smart response to becoming indispensable is to embrace the regulation it summons rather than fight it, you are looking at something structural. Becoming essential and keeping unlimited pricing power are incompatible, and the cleverest men in these industries understood it and made their peace with it in advance. The regulatory compact that Insull championed, protected territories in exchange for capped returns and universal service, became the template for the American electric utility and formal state regulation of utilities, traced commonly to Wisconsin in 1907 and spread across the country on the model he had argued for.

The owners of the grid did fine under this arrangement. They did not go broke the way the railroads did. But they did not get rich in proportion to how essential their product became, either, because the moment electricity became truly indispensable was the same moment its owners lost the right to price it like the miracle it was. They became utilities: stable, protected, permanent, and capped. That raises the question this book keeps asking. If the owners of the grid were capped, where did the real fortune go?

Where the Fortune Went

It went, first and most visibly, to General Electric, and GE is worth pausing on because it is the ancestor of a company type that matters enormously for the AI argument. GE was formed in that 1892 Morgan merger, folding Edison's manufacturing operations into a new giant, and it did not merely build the dynamos that powered the grid. It built the generators and the appliances and the industrial motors and the light bulbs and the turbines, the infrastructure and the things that ran on the infrastructure, all under one roof. It owned the grid layer and the build-on-top layer at the same time. It was not the regulated utility, capped and territorial. It was the company that sold the utility its equipment and sold the public the machines that made them want electricity in the first place. GE grew into one of the largest and most successful corporations in the history of the world. Remember GE, because when we get to the artificial intelligence chapters, the most powerful companies in the story are going to look far more like GE, which owned both layers, than like Commonwealth Edison, which owned only the wires.

But the deepest lesson about where the value went is subtler than GE, and an economist named Paul David explained it in a way that reframes the whole history. When electric motors became available to factories, they did not immediately improve productivity. For roughly forty years the promised gains simply failed to appear. The reason, David showed, was that factory owners installed the new electric motors inside factories still designed for the old technology. A steam-powered factory was built around a single massive central engine, which drove every machine in the building through an elaborate system of shafts, belts, and pulleys running along the ceilings. When electricity arrived, owners at first just replaced the big steam engine with one big electric motor and kept the shafts and belts. They had electrified the factory without rethinking it, and the gains were meager.

The productivity revolution came only when a new generation of manufacturers threw out the whole design. Instead of one giant motor driving everything through belts, they put a small electric motor on each individual machine, a system called unit drive, so that every workstation had its own independent power and could be placed anywhere, run at its own speed, and switched on and off by itself. This freed the factory floor from the tyranny of the central shaft. Machines could be arranged in the logical order of the work rather than the order dictated by the belt system. The factory could be lit by windows instead of darkened by overhead shafting. And once factories were rebuilt around what electricity actually made possible, productivity exploded. The manufacturing boom of the 1920s, the surge that built the modern American economy, was in large part this reorganization finally paying off, decades after the electric motor itself was old news.

The single greatest monument to this is Henry Ford's assembly line, which began moving at his Highland Park plant in October of 1913. Ford did not simply add power to car-making; he rebuilt the entire process around the assumption of abundant, flexible, electrically driven machinery, breaking assembly into dozens of sequential steps and moving the work to the worker. The result cut the time to assemble a Model T from more than twelve hours to about ninety minutes, and the price of the car from more than $800 to around $300, putting an automobile within reach of the very workers who built it. The fortune Ford built was not a fortune in generating electricity. It was a fortune in reorganizing an industry around cheap electricity as a free input.

And beyond Ford, the cheap power Insull and his kind produced created an entire consumer economy that had never existed. The electric range, the refrigerator, the washing machine, the vacuum cleaner, the radio, a whole world of household appliances, each of them a business built on top of the assumption that electricity was cheap and always available. The streetcar suburbs, made possible by the electric interurban lines that Insull himself owned, filled with new houses wired from the start for this new way of living. None of these industries generated a watt of power. They were built on top of the grid, and they were where the durable fortunes of the electrical age accumulated. Insull made electricity cheap. Other people got rich on what cheap electricity made possible.

This is the lesson aimed directly at the reader of this book, and it is Paul

David's lesson brought forward a century. The fortunes in the age of artificial intelligence will not go, in the main, to the companies that generate intelligence, any more than they went to the companies that generated electricity. They will go to the people who reorganize their businesses, their industries, their whole way of working, around the assumption that intelligence is cheap and everywhere. The same way Ford reorganized manufacturing around cheap power. Bolting a language model onto an unchanged process is the equivalent of replacing the steam engine with one big electric motor and keeping the belts. The surplus is hiding in the reorganization, and the reorganization is hard, slow, and exactly where the value lives.

The Fall

Samuel Insull should have ended his life as a revered elder statesman of American industry, the man who electrified the Midwest and invented the modern utility. Instead he became, for a time, the most hated man in America, and his story took a turn so dark and so operatic that no novelist would risk it.

The instrument of his destruction was the same financial cleverness that had built his empire turned against him. To finance his endless expansion, Insull had built, on top of his sound and profitable operating utilities, a towering structure of holding companies, corporations that existed only to own shares in other corporations, stacked on top of one another in precarious layers of debt and cross-ownership. As long as the whole edifice kept rising, the leverage magnified his wealth and his control magnificently. But leverage runs in both directions, and Insull had built a machine exquisitely sensitive to any reversal.

The reversal was the Great Depression. When the stock market collapsed in 1929 and the economy fell into ruin, the value of the holding companies at the top of Insull's pyramid evaporated, and the debt that had leveraged them became a noose.

It is crucial to understand, however, because it is the heart of the matter, that the actual electric utilities at the bottom of the structure, Commonwealth Edison and Peoples Gas and the rest, remained sound and profitable and kept paying their dividends straight through the Depression. The lights never went out. The business of generating and selling electricity was as solid as ever. What failed was the tower of financial engineering Insull had built above that business, the holding companies leveraged against holding companies, and when it began to fall, it fell with astonishing speed. His empire had taken fifty years to build. It came apart in roughly fifty days.

In April of 1932, Insull's Middle West Utilities went into receivership. He tried to save the situation with a $10 million loan, and a group of New York bankers, old rivals, moved to block it. The loan did not come. A $3 billion empire was forfeited for want of $10 million. In June the seventy-two-year-old Insull resigned everything, some sixty chairmanships and eighty-five directorships and eleven company presidencies. He sat for hours at his desk signing away the work of his life until he was left with almost nothing.

Then the country turned on him, and the turn was ferocious. The man who had been celebrated as a business genius became, overnight, the symbol of everything that had gone wrong, the face of the greedy financier whose recklessness had ruined ordinary investors. Small shareholders who had trusted the Insull name and lost their savings wanted a villain, and the collapsing economy wanted one even more. Insull was cast in the role. In the presidential campaign of 1932, Franklin D. Roosevelt denounced him by name, thundering about "the Ishmael or Insull, whose hand is against every man's." Federal and state prosecutors, feeling the same political current, began assembling criminal cases against him for mail fraud, embezzlement, and violations of the bankruptcy laws.

Fearing that the political fury made a fair trial impossible, Insull fled the country before the indictments came down, first to France and then to Greece, which had no extradition treaty with the United States. The flight only inflamed the coverage, turning him into an international fugitive on the front pages. When American pressure grew too great, he was seized by Turkish authorities and returned under guard to stand trial in Chicago.

He was tried three times between 1934 and 1935 on federal charges of mail fraud, state charges of embezzlement, and federal charges of violating bankruptcy laws. He was acquitted every single time. In the first and most important trial, the jury took just over two hours to conclude he was not guilty on all counts. It was said afterward that they had made up their minds almost at once and waited only so that no one could accuse them of having been bought. The juries could see what the newspapers and the politicians would not admit, that a man could preside over a catastrophe without having committed a crime.

That the collapse had been caused by a global depression and a structure of legal financial engineering rather than by fraud, and that no single person could possibly have caused a disaster of that magnitude. Insull had lost everything, but he was not a criminal, and twelve of his fellow citizens, three times over, refused to pretend otherwise.

It did not restore him. His reputation was destroyed, his fortune was gone, and he spent his last years abroad, subsisting on a modest pension in Paris. On a summer day in 1938, he collapsed of a heart attack on the stairs of a Paris Métro station and died there, largely alone. The story that he was found with only a few cents in his pocket may be embellished, but the shape of the ending is true. The man who had controlled billions and electrified a third of the nation died nearly broke on the steps of a subway in a foreign country, an object lesson in how fast and how far a great fortune can fall.

There is a bitter irony in the aftermath. The public revulsion at Insull's collapse helped drive the passage, in 1935, of the Public Utility Holding Company Act. A law designed specifically to outlaw the kind of leveraged holding company pyramids Insull had built. His lifelong vision, that electric power should reach everyone, even the unprofitable countryside, found its fulfillment in the same New Deal that had made him a scapegoat. The man was destroyed and his ideas were enshrined, which is its own kind of epitaph.

The Countryside

That vision, of power for everyone, brings us to the last piece of the electrification story, and it is the piece that completes a thread running through this whole book.

For all of Insull's genius at driving electricity cheap and wide, there was one place the private power companies would not go, the American countryside. The economics that made city electrification so profitable ran exactly backward on the farm. Rural customers were few and far apart, the cost of stringing wire to reach them was enormous. The revenue from a scattering of farmhouses could never justify it. So the private utilities, Insull's included, simply declined. As late as the mid-1930s, when American cities had been brightly lit for a generation, the great majority of American farms still had no electricity at all. Farm families cooked and cleaned and did their books by the same lamplight their grandparents had used. A country that had electrified everything profitable stopped precisely at the edge of profit.

This is the same wall we hit in the last chapter, when the telephone reached the profitable places and needed a system of cross-subsidies to reach the rest. With electricity, the market failure was even starker, and the solution was more direct. The market would not wire rural America, so the government did it. Franklin Roosevelt created the Rural Electrification Administration (REA) by executive order in 1935, and Congress made it permanent with the Rural Electrification Act of 1936. Rather than trying to force the reluctant private utilities to serve the farms, the program financed rural families to build their own electric cooperatives. Member-owned nonprofits that would string the lines the companies wouldn't and run the power themselves. It worked, and it worked fast. Within about fifteen years, rural America went from mostly dark to mostly wired, one of the most successful public infrastructure programs in the nation's history.

The lesson is one you've now watched play out twice. I'm stating it here because it is going to matter enormously at the end. When a technology becomes truly foundational, when it crosses from a convenience into a necessity for full participation in modern life, the market alone will not deliver it to everyone. The market delivers it to the profitable and stops. When a society decides that universal access matters more than the logic of profit, it reaches in and finishes the job the market abandoned. It did so by the telephone's cross-subsidies and, here, by the countryside's electric cooperatives. The guarantee of access is the other half of the bargain that caps the owner's returns, and the two halves are inseparable.

We watched telephone service reach the farm by one mechanism and electricity reach it by another. When we reach the question whether capable artificial intelligence will be available to everyone or only to those a market finds profitable to serve, this is the history sitting behind it. The question of an "REA for AI" will not seem nearly as far-fetched as it first sounds.

What Insull Understood

Viewing the whole arc of this story, the six questions I put to every case answer themselves once again, with electrification giving the fullest answers yet.

How did it cross from novelty to necessity? It crossed through the War of the Currents and the victory of alternating current, and then through Insull's flywheel of ever-cheaper power, until a nation that had lived by lamplight could not imagine living without the grid.

Who paid to build it, and did they keep the wealth? Insull built the machine and chose the regulated cap, and the durable fortunes went not to him but to General Electric and to the reorganizers like Ford. Insull himself was ruined by the leverage he had stacked on top of his sound utilities.

Did it commoditize and get regulated? It did, and this time, as with the telephone, the regulation was chosen by the industry's leading figure rather than forced upon it.

Where did the value go? Up the stack to the company that owned both layers and to everyone who rebuilt their business around cheap power.

And what did people fear? Here, uniquely among our cases so far, they feared the technology itself, feared that it would kill them, and the fear was real. Despite this fear, electricity survived and thrived.

Four lessons carry forward to artificial intelligence, and they are the sharpest this book has yet assembled.

The first is that the trap was chosen, yet again, and being chosen twice makes it law. Vail did it with the telephone and Insull did it with electricity, independently, in the same era. Two of the age's great industrialists reaching the same counterintuitive conclusion that the response to indispensability is to embrace regulation rather than resist it. Watch for the company in the AI build-out that does what Insull and Vail did, that reads the trap coming and walks in on purpose. It will look, at first, like a company acting against its own interest. It will actually be the smartest company in the room.

The second is that commoditization is a strategy, not a catastrophe. Insull got rich by making electricity cheap, deliberately, relentlessly. He understood that the fortune was in owning a huge share of an enormous cheap flow, not a small share of an expensive one. The people who look at the collapsing price of machine intelligence and see doom for the companies building the grid are making the error Insull never made. Cheap was the plan. If the collapsing price is Jevons paradox all over again, it will not shrink the demand for intelligence. It will explode it, exactly as cheap power did not shrink the appetite for electricity but created the entire electrified world.

The third is Paul David's lesson, and it is aimed straight at the builder reading this. The fortunes did not go to the dynamo owners. It went to the reorganizers, to the manufacturers who tore out the steam-era layout and rebuilt around unit drive. It went to Ford, who rebuilt an industry around cheap power. In the AI age, the surplus will hide in exactly the same place. In the hard, slow work of rebuilding a business around the assumption of abundant intelligence, not in bolting a model onto a process designed for a world without one.

And the fourth is the one that separates this chapter from the rest and points furthest ahead. The fear of the technology itself, the safety fear. It was real in the War of the Currents, and it was survived. Electricity did kill people, and society did not respond by rejecting it or pretending it was safe, but by doing the patient work of making it safe enough. That was accomplished through standards and regulation and rulemaking so that its benefits could be captured. Whether the safety fear about artificial intelligence follows a similar path is an open question. Is it a real danger that can be engineered and regulated down to an acceptable level while the technology transforms the world? Or is it different in kind, a danger that does not yield to the old methods? This is the question I will end on. Electrification does not answer it, but it is the reason the question is worth taking seriously in both directions, against the boosters who say there is nothing to fear and against the doomsayers who say the fear is the end of the story.

Samuel Insull understood almost everything about his industry. He understood that the fortune was in cheap and abundant rather than scarce and dear. He understood that an indispensable monopoly had to accept regulation to survive. He understood, before almost anyone, that electricity would remake the whole of modern life. What he did not understand, or understood too late, was that a man can build something magnificent and essential and still lose it all. That becoming indispensable protects the thing you built far better than it protects you. The grid he assembled still hums under the Midwest today. The company still sends out its bills. And almost no one who pays them has ever heard the name of the small, precise, doomed Englishman who made electricity cheap enough to take for granted.

Chapter 6

The Roads Were Already Built

> If you build it, they will come.
>
> —the animating faith of the fiber boom, as *Wired* described Qwest's strategy in 1998

The latter half of the 1990s and the first years of the new century are remembered as the dot-com era. It was a time of profound technical change and innovation. The internet burst onto the scene and transformed how business was done, how people lived their lives, and how the world communicated. It was the true birth of the digital age.

In reality, two separate technology grids emerged at the same time. The first was the internet itself, which for most people meant the World Wide Web, the .com in dot-com. Early on, though, it was just as often the world wide wait, as pages crawled onto the screen through dial-up modems and twisted copper wire.

The second grid was the one that ignited the first. Fiber optic lines sped the internet up by a factor that is hard to overstate. A standard dial-up modem of that vintage ran between 300 baud and 56 kilobits per second. Modern residential fiber routinely delivers anywhere from 300 megabits to 10 gigabits per second, millions of times faster.

That fiber grid is the fourth great build-out of this story, and the subject of this chapter. The collapse in the price of bandwidth as fiber lost its scarcity caught some of the smartest businessmen of the era flat-footed, and I was standing next to a few of them when it happened.

My own part in all this was minor in the arc of history. I cofounded two companies and helped build a third in those years, and they shaped who I am and how I view the world. One of them was Interliant, the company the Prologue found in that room in Houston. We were what was called an ASP, an application service provider, by then anyway. We had started off as a roll-up in the web-hosting business, offering the three P's: power, ping, and pipe. That style of pure hosting was commoditizing rapidly. We had tried to move up the stack by offering "fully managed hosting," but that was a bit of a ruse; it just made us power, ping, pipe, and fingers. We kept going further up the stack, to offer hosted enterprise applications. To give you some perspective, this is pre-cloud, pre-virtualization.

Interliant had data centers across the US and Europe, and one of our biggest costs was bandwidth. Just over a year earlier, we had signed what we thought was a great deal with MCI WorldCom, Bernie Ebbers's company. It let us bring in about as many DS3 lines, the Cadillac of connectivity at the time, 45 megabits per second of bandwidth each, as we could use and resell. The price worked out to something like $1,200 a month per megabit per second, somewhere north of $50,000 a month for a single line. I had helped get the deal done.

Within a year of signing, the street price of that same bandwidth had collapsed toward $600 or $700, and it was still falling. A dozen or more companies were racing to bury fiber-optic cable across the country faster than anyone could possibly use it, and the dark fiber (the fiber laid but waiting for traffic to come) was piling up. So when a company got in touch to say it was setting up a commodity-style bandwidth trading operation, one that would let us swap our high-priced bandwidth for equal-quality bandwidth at competitive street pricing, our CEO, our COO, and I flew down to Houston to meet with them.

The stock market, at that moment, was having a little blip. We were sure it was momentary, a minor correction, and that things would be back up and to the right shortly. Little did we know that the dot-com bubble was bursting around us, and that both the company we were buying the bandwidth from and the company we had just met with would end up embroiled in criminal-level malfeasance. The company in Houston, as the Prologue revealed, was Enron, and you know how that story ends.

I opened the book with that scene not because I have a war story to tell, though I sure have a few, but because that room in Houston contained, in miniature, the entire subject of this chapter. Enron was trying to trade bandwidth the way it traded natural gas and electricity, to turn the capacity of a fiber-optic network into a commodity you could buy and sell in standardized contracts with what they called "pooling points" that were, in their own words, to bandwidth what Henry Hub is to gas. The reason the whole thing had the flickering unreality of a con was that the underlying good was collapsing in value so fast that no amount of financial cleverness could keep up with it. Bandwidth was becoming too cheap to trade profitably even as Enron built a business on trading it. The commodity was commoditizing in real time, right under the feet of the people trying to corner it.

This is the most recent of the four build-outs in this book, the one I lived through as a working participant rather than read about in a history book, and it is the one that rhymes with the artificial intelligence build-out most precisely, because it is the closest in time and the freshest in memory. In fact, much of the AI build-out is being built on the bones of the dot-com era.

It has all the elements we have seen before. A frenzy of overbuilding financed by borrowed money. A spectacular crash. A technology that became indispensable and then had its owners disciplined by the state. Above all the same migration we have watched in every chapter, the value climbing off the infrastructure and up to the people who built on top of it.

The Glut

To understand why the bandwidth in that Houston conference room was collapsing in value, you have to understand what had just happened to the American landscape. One could say truthfully that it was being carpeted in glass. A type of glass rolled into a tube thinner than a human hair, that allowed light pulses to travel through at extremely high speeds and changed the world.

The trigger for all this was a law. The Telecommunications Act of 1996 deregulated the long-distance market, opened the internet backbone to private competition, and effectively invited anyone with capital to go build a national network and compete with the incumbents. Wall Street answered the invitation with something close to hysteria.

This was the dot-com boom, the internet was going to change everything, and the animating belief of the moment was that whoever controlled the physical pipes through which all those bits would flow would own the future. Internet traffic, everyone repeated to everyone else, was doubling every hundred days. The figure was mostly false, but it was said so often and so confidently that it became the foundation for hundreds of billions of dollars in investment. *Wired* magazine captured the faith in 1998, describing Qwest's strategy in a single phrase: If you build it, they will come.

So they built and built some more. They raised money to keep building. Let us pause a moment on the companies who did the building, because they were, in a very literal sense, the descendants of the railroads. Qwest, one of the largest of the new carriers, had grown out of a railroad construction company and strung its fiber along the rail corridors that ran from Denver to the coast, laying the newest infrastructure of the information age in the roadbed of the oldest infrastructure of the industrial one. Williams Communications threaded its cable through the rights-of-way of natural gas pipelines. Perhaps best of all, Sprint, the name itself, was an acronym left over from its origins inside a railroad: Southern Pacific Railroad Internal Networking Telecommunications.

The path of the fiber followed the path of the rails and the pipelines because the rights-of-way had already been cleared a century before. That is the first and most literal instance of this chapter's title. Before anyone spoke of the internet running on roads that were already built, the fiber ran on roadbeds that were already graded.

The scale of the building was staggering, and it was almost entirely financed with debt. In the five years after the 1996 Act, American carriers spent more than $500 billion, and by some estimates the industry took on close to a trillion dollars in debt, laying something on the order of 80 to 90 million miles of fiber-optic cable.

Bond markets and equity investors poured money into any company with a plausible plan to lay glass. Equipment makers like Lucent and Nortel and Cisco lent the carriers the money to buy their own equipment, a practice called vendor financing that booked the loans as revenue and flattered everyone's numbers on the way up. It was a self-reinforcing machine of optimism. Rising stock prices made it easier to raise capital, and fresh capital pushed valuations higher still.

And then the demand did not show up. Or rather, it showed up eventually, years later, but nothing fast enough to justify what had been built. When the accounting was done, it turned out that only something between 5 and 10 percent of the fiber laid during the boom was ever actually "lit," ever carried a single bit of traffic. The rest sat in the ground, unused, and the industry gave it a name that has a certain poetry to it. They called it dark fiber. Ninety percent of the greatest communications build-out in history was dark, humming with nothing, waiting.

The reckoning, when it came, was not a grind. It was a crash, and it was among the worst in the history of American business. Between 2000 and 2002, telecommunications companies lost more than $2 trillion in market value. Global Crossing, which had laid more than a hundred thousand miles of undersea and terrestrial cable across twenty-seven countries on a vision of connecting the world, went bankrupt in January of 2002 owing more than $12 billion. That same summer WorldCom, which had grown by acquisition into one of the largest telecommunications companies on earth, filed what was then the largest bankruptcy in the history of the United States, buried under some $41 billion in debt and exposed as the site of a nearly $4 billion accounting fraud. The industry as a whole was left holding roughly a trillion dollars in debt that would never be repaid. Corning, which made the glass, saw its stock fall from over a hundred dollars to about two. Companies that had employed tens of thousands laid off most of them. Hundreds of thousands of telecom jobs vanished.

I want to be precise about why this was a crash and not the gentler kind of correction we will see in a moment with the cloud, because that distinction is extremely useful in regard to our own moment. The fiber build-out crashed because of how it was financed. It was built by pure-play companies, firms that did nothing but lay fiber, using money borrowed from bond markets and banks and vendors, money that had to be paid back on a schedule regardless of whether the demand had arrived. When the traffic failed to materialize on time, there was no cushion. There were no other profitable businesses to carry the losses, no patient internal cash flow to absorb the shortfall, only debt service coming due against revenue that was not there. Leverage is a magnificent engine on the way up and a guillotine on the way down, and the fiber industry had strapped itself to the blade. Keep in mind the connection between outside debt and the violence of the crash, because when we reach the artificial intelligence build-out and ask whether it ends in a bang or a slow bleed, the answer is going to turn on exactly this question of who is paying for it and with what.

The View from Inside

I was not a spectator to any of this. I was a customer of it, and a participant in it, and my experience turned out to be a compressed demonstration of the argument this entire book is making. Since I lived it, let me use it, with the warning that I am telling it not to reminisce but because it illustrates the mechanism more cleanly than any example I could borrow from news clippings and history books.

My first company, TriStar Web, started out building websites. This was the mid-1990s, the very beginning of the commercial web, and everyone with a business suddenly needed a presence on it. A whole industry sprang up to build those presences. It was the dawning of the age of the web developer. The glamorous end of that industry was occupied by firms like Razorfish and Agency.com, the celebrated web-design shops of the era, and we competed down in the trenches beneath them. What we learned, building website after website, was that it was a miserable business. Not because the work was hard, though it was, but because it was the wrong layer. You would spend weeks on a site and then spend weeks more arguing with the client about this shade of blue versus that one, about whether the oval should be a little more oval. It was custom, artisanal, low-margin work, and every project started from zero, and there was no way to build durable wealth doing it because you were selling your hands and your taste by the hour. It was, in the language of Chapter 2, the tip of the stack, the place where human labor lives and the margins are thin.

So we made a move, made by instinct, before I had any framework to describe it. We stopped building websites and started hosting them. Instead of making the site, we would house it, run it on our servers, keep it up and available, and charge a predictable monthly fee for the service. We became, in effect, digital landlords. It worked beyond anything the design business had ever offered. We charged something like $50 a month per site, but the beauty of it was that it recurred, and it scaled, and it did not require us to argue about ovals. Within eighteen months we were hosting something like five thousand websites. We had climbed down off the artisanal top layer and onto the infrastructure below it, and the infrastructure was where the money was, exactly as every chapter so far has insisted.

Then the thing that happens to everyone in this book happened to us. The layer we had climbed onto began, itself, to commoditize. Hosting a website was not hard, the barriers to entry were low, and competitors multiplied, and the price of hosting began the same slide toward the floor that the price of bandwidth was soon to be on. We did what companies do when the ground is eroding beneath them. We tried to get bigger and to climb higher at the same time.

We sold TriStar Web to a company called Sage Networks, which was doing a roll-up in hosting, and after the acquisition I became a VP of business development. We were one of several firms rolling up the industry, buying something like thirty companies in thirty-six months, taking strategic investment from the likes of Dell, Microsoft, and IBM, and eventually going public in the great dot-com offering wave under the name Interliant. As things kept commoditizing, we tried to climb up the stack again. We pivoted from merely hosting websites to hosting applications. We were running actual business software for our customers over the wire, a business model that in those days was called an ASP, an application service provider.

We were right about where the value was going, and we were early by exactly one technological generation. Unfortunately for us, being early by one generation is indistinguishable, in its financial consequences, from being wrong. The application-hosting business we were trying to build is essentially the cloud. It is the thing that Amazon, Microsoft, and Google would later build into the most profitable enterprise in the history of technology.

But we could not make it work in 2000, and the reason we could not is worth stating precisely, because it is the hinge of this chapter. Bandwidth was still too expensive, though it was falling fast. But more fundamentally, the essential invention that makes the cloud economically possible did not yet exist. We were putting one customer on one physical server, because that was the only way anyone knew how to do it, which meant our costs scaled linearly with our customers and we could never get ahead. The thing that would eventually break that constraint, that would let one physical machine safely and efficiently serve many customers at once, was a piece of software called the hypervisor, and in the years we needed it, it was not ready.

So Interliant did not survive. It was caught in the shakeout, another company that had bet on infrastructure at the moment infrastructure was collapsing in value, and the wave that carried Google and Amazon to the heights washed us under. I tell you this not for sympathy, which no entrepreneur deserves for a bet that didn't pay off, but because I was living, without the vocabulary to name it, the precise lesson of the chapter before this one.

In the electrification chapter, we watched the economist Paul David explain that the electric motor did not transform the factory until the factory was reorganized around it. Until unit drive replaced the central steam engine and the belts, and that the reorganization took forty years. That the fortunes went to the reorganizers who moved at the right time and not to the ones who moved too early.

I was one of the ones who moved too early. The hypervisor was the unit drive of the cloud, the enabling reorganization that had to arrive before the value could be captured. Interliant tried to build the electrified factory before the unit-drive motor had been invented. Being right about the destination is worth nothing if the road to it has not yet been built.

The Machine That Made the Cloud

The hypervisor deserves its own moment, because it is the invention that turns this chapter from a story about a crash into a story about what grew out of the crash.

For all of computing history up to this point, a physical server ran one operating system and did one set of jobs, and it ran them badly in one specific sense: It sat mostly idle. Studies of the era found that the typical server ran at something like 10 or 15 percent of its capacity, which meant that companies were buying enormously expensive machines and using a small fraction of what they had paid for, and there was no good way around it. You could not safely put multiple customers or multiple critical applications on the same machine. If they shared an operating system, they could crash into one another, corrupt one another, see one another's data.

This is, if you have been reading closely, precisely Samuel Insull's problem from the last chapter transposed into silicon. Insull's power plants, built to meet the evening peak, sat half-idle the rest of the day, and his entire fortune came from the techniques he invented to fill that idle capacity, to run the expensive plant closer to full by mixing customers whose demands came at different times. The idle server is the idle power plant. And the hypervisor is the tool that finally solved it, seventy years later, for computing.

A hypervisor is a thin layer of software that sits between the physical machine and the operating systems running on it, and it does something that sounds modest and is actually revolutionary. It lets one physical computer pretend to be many separate virtual computers, each fully isolated from the others, each believing it has the machine to itself. Suddenly you could safely pack many customers onto one server, run it near full, and give each customer the clean illusion of their own private machine. The economics of computing inverted. The idle plant could finally be filled.

The company that made this practical was VMware, which shipped the first commercially successful virtualization of the standard x86 computer in 1999 and its server version, ESX, in 2001. There is a cruel irony in those dates for me personally, because they bracket precisely the window in which my own application-hosting business was failing for want of exactly this technology.

The full economic power of virtualization took a few more years to mature, arriving properly when Intel and AMD built support for it directly into their chips in 2005 and 2006. And it is not a coincidence that the modern cloud was born in exactly those years, because the cloud is what you get when you point industrial-scale virtualization at the problem of selling computing to the masses.

In March of 2006 Amazon launched a service called S3, for storing data, and that August it launched EC2, which let anyone with a credit card rent a virtual computer by the hour and give it back when they were done. This was the cloud, and it changed the physics of building a company. Before it, starting an internet business meant raising money to buy servers, exactly the trap my company had been caught in. After it, you rented what you needed, paid only for what you used, and scaled up or down in minutes. The capital barrier that had defined the entire dot-com era simply dissolved.

Now there is a tidy story about where AWS came from. Urban myth perhaps, but it goes something like that Amazon had built enormous computing capacity for its own retail business, which peaked at Christmas and sat idle the rest of the year. They were trying to rent out that excess. The idle plant model yet again. That story is partly true and partly another tech-industry legend that has grown in the retelling.

At least some of the people who built AWS insist it was purpose-built rather than repurposed. I will not lean too hard on the tidy version. But the underlying economic logic holds regardless of the origin story, and it is Insull's logic exactly: Computing is a hugely expensive plant that is ruinous when it sits idle and enormously profitable when it runs full, and the entire cloud business is a set of techniques, beginning with the hypervisor, for filling the plant.

Here is the thing I most want you to notice about the cloud, the thing that makes it the perfect companion to the fiber crash that preceded it. The cloud was also an enormous, speculative, overbuilt infrastructure. Amazon and Microsoft and Google poured tens and eventually hundreds of billions of dollars into data centers, far ahead of proven demand, on faith that the demand would come. In other words, they did essentially the same thing the fiber companies had done. And yet the cloud did not crash. There was no cloud equivalent of the 2001 telecom collapse, no wave of hyperscaler bankruptcies, no two-trillion-dollar evaporation. The build-out was absorbed, digested, grown into over a decade of steady expansion. Why?

The answer is the one I asked you to hold earlier, and it is the whole of the matter. The difference was the financing. The fiber companies were pure plays built on outside debt that had to be serviced on a schedule. The cloud was built by three of the most profitable companies in the history of capitalism, out of their own cash flow. Amazon's retail machine, Microsoft's software licenses, Google's advertising empire, these threw off rivers of cash that could fund a decade of speculative data-center construction without a single anxious conversation with a bondholder.

When you build an overbuilt infrastructure with borrowed money, you get the fiber crash. When you build the same overbuilt infrastructure with your own overflowing profits, you get the cloud, a long, absorbed, undramatic grind up and to the right. Same overbuilding. Same speculation about future demand. Opposite financing. Opposite outcome. There is no more important lesson for anyone trying to guess how the artificial intelligence build-out ends, and we will return to it in earnest when we get there.

Where the Fortune Went, Again

Back to the fiber. It was laid, and a lot of it stayed dark. The layers were stacked, and now we ask the same question we have asked of every build-out, and here we get the sharpest answer yet. Where did the value go?

It did not stay with the people who laid the fiber. That much is beyond dispute, because most of them went bankrupt, or worse, to jail. Global Crossing, gone.

WorldCom, gone, its chief executive imprisoned. Qwest, engulfed in its own accounting scandal. Enron, of course, gone in the most spectacular corporate collapse of the age. The men who had borrowed a trillion dollars to carpet the country in glass, who had believed that whoever owned the pipes would own the future, were, within a few years, the greatest losers in the whole affair. Owning the pipes turned out to be the worst place to be.

The value went up the stack, to the companies that treated all that fiber as a free input. When Global Crossing's hundred thousand miles of cable and all the rest of the dark fiber were sold off in bankruptcy for pennies on the dollar, they did not disappear. They became the cheap, abundant backbone of the internet for the next twenty years, and the companies that built on top of that suddenly-cheap backbone inherited the earth. Google, founded in 1998, in the very teeth of the boom, built a business that assumed connectivity was essentially free and infinite, and became one of the most valuable companies ever to exist. Amazon did the same, and then went further and became the landlord of the whole new layer through AWS. Netflix mailed DVDs until the bandwidth got cheap enough to stream, and then rode the dark fiber into every living room in the world. An entire economy of software companies, the businesses we came to call SaaS (software as a service), grew up assuming the cloud beneath them the way you assume the electrical outlet in the wall, and never gave a thought to who had gone bankrupt laying the cable that made it all possible.

This is the same shape we have now seen four times, and for the people in the back, let's say the four aloud, because the repetition is the proof. The railroads went bankrupt and Carnegie and Rockefeller and Sears got rich building on the rails. The telephone monopoly was regulated and broken up while the transistor it invented, given away by government order, built Silicon Valley on top of it. Insull was ruined while General Electric and Henry Ford and the appliance economy got rich on the cheap power he had made. And now the fiber carriers went bankrupt while Google, Microsoft, and Amazon got richer on the cheap bandwidth they had stranded. Four build-outs, four sets of infrastructure owners capped or bankrupted or broken, and four times the durable fortune migrated up and off the infrastructure to the people who built on top of it. If you are looking for the single most reliable pattern in the economic history of technology, this is a strong candidate, and it is the reason this book exists.

The Company That Refused to Stay in Its Lane

But I have to stop here and confront the strongest objection to everything I have just argued, because it comes precisely from the companies I just named as the winners. Google, Microsoft, and Amazon did not merely build on top of the cheap infrastructure. They built the next infrastructure themselves. Amazon took the profits from riding the cheap bandwidth and poured them into becoming the landlord of the cloud, the very layer beneath the next generation of builders. Google did the same. Microsoft got there by a different road, taking the profits from its stranglehold on the desktop operating system and the office suite and buying and bullying its way into the cloud game. To do it the company had to swallow something that would once have been unthinkable. Microsoft had to run Linux, the open-source operating system it had spent a decade treating as the enemy. That was a bitter pill for a company built on proprietary Windows, but it was the price of a seat at the hyperscaler table, and Microsoft paid it.

These are not companies that stayed politely up the stack and let someone else own the grid. They own the grid and the layer above it and the layer above that, all at once, and they are among the most profitable enterprises in the history of the world. Does that not break the thesis? If the same company can own the commoditizing infrastructure and also capture the value that migrates up off it, then the value has not migrated away from the infrastructure owner at all. It has simply migrated into a different pocket of the same enormous coat.

This is the most serious challenge in the book but meeting it honestly and head-on makes the argument stronger rather than weaker. The answer is already sitting in the last chapter, in a company I told you to remember. The answer is General Electric. GE never was the regulated utility, capped and territorial, stuck owning only the wires. GE owned the dynamos and the appliances and the motors and the light bulbs, the infrastructure and the things that ran on it, and it grew colossal precisely because it refused to stay in one layer. It was the defining conglomerate of the industrial age. It was the supplier to the utility, not the utility itself. The thesis of this book was never that infrastructure owners always lose. The railroads lost, and the fiber carriers lost, and Insull's holding companies lost because they were pure plays, trapped at the bottom of the stack owning nothing but the commoditizing layer. The thesis is subtler and more useful than "infrastructure loses."

It is that value migrates up the stack, relentlessly, and the only infrastructure owner who keeps it is the one who climbs up the stack along with it. The pure play gets commoditized and stranded. The integrated giant who owns the grid and follows the value upward becomes GE. The hyperscalers are not a refutation of the pattern. They are the modern GE, and they are the proof that the way to survive the trap is to refuse to stay in the lane the trap assigns you. It is exactly what we tried to do at Interliant, and failed, because we were about one technological generation too early.

And here the story gets genuinely interesting, because these companies have already run this exact play once, successfully, and are now facing the decision to run it again one rung higher. Watch the sequence. The cloud began as raw infrastructure, what the industry calls IaaS (infrastructure as a service), the digital-landlord business of renting out bare compute and storage and bandwidth. That layer commoditized, exactly as the pattern predicts, into a brutal price war where a gigabyte of storage races toward free.

So the hyperscalers climbed. They moved up to what is called PaaS (platform as a service), the managed databases and queues and operational plumbing that are stickier and higher-margin and harder to leave. And that PaaS layer became the foundation on which the entire modern software economy, the SaaS companies, built their businesses. The hyperscalers had done the GE move in miniature and in fast motion. They owned the commoditizing bottom layer, and rather than get stranded on it, they climbed to the layer above and captured the value the builders on top were throwing off. From their vantage, this migration is not a threat. It is a thing they have already beaten once.

Now artificial intelligence arrives, and it commoditizes the PaaS perch the way PaaS commoditized the raw infrastructure beneath it, and the hyperscalers face the same fork one rung higher up the stack. And the fork is exactly the set of questions the chapters ahead have to answer.

Do they want to be AI factories, buying the GPUs by the hundred thousand and building the mega-clusters to train and serve models, competing head to head with a new crop of specialized rivals, the neoclouds and the GPU-for-rent shops that have sprung up to do nothing else? Do they want to be the inference engines that serve the intelligence to everyone else's applications? Or do they want to be the platform that others build their AI on top of? Because that last choice contains a trap that should by now be familiar.

If a hyperscaler decides its role is to let everyone build on top of it, then it has volunteered to be the grid. It has chosen to be the Commonwealth Edison of artificial intelligence, the essential, capital-devouring, regulated-in-all-but-name layer that everyone depends on and no one overpays, while the value climbs past it to the applications above. The GPU-rental business, stripped of its branding, is the fiber-carrier business wearing a better suit: enormous capital expenditure, savage competition, thin margins, and the fortune migrating up and off to whoever builds on the cheap compute.

There is a deeper version of the fork, and it is the one that could matter the most. Do they want to own the model itself? To climb the final rung, from owning the grid to owning the intelligence that runs on it? And here, for the first time in four build-outs, the great companies are visibly disagreeing with one another, placing different bets on the same table, which means the next few years are going to run a natural experiment of a kind economic history rarely offers. Some are climbing all the way to the model and some are deliberately declining to, and the pattern traced makes a different prediction for each of them. That disagreement, and what it tells us about where the value in artificial intelligence is finally going to come to rest, is the subject of the chapters ahead.

But before we get there, let me add the twist that unsettles even the companies climbing fastest, the possibility that makes this harder than a simple instruction to be GE and own the model. It is this. Even the model may not be a safe place to stand, because the model itself is commoditizing.

The intelligence at the base of all of this, the raw capability of a frontier model, is not going to keep pulling away from the pack the way it did in the first few years. The gap between the best model and the fifth-best is collapsing. The benchmarks are bunching up near the top. The gains from the old recipe of more data and more compute are getting smaller with each turn, because the data to train on is running out and each doubling of computing power buys a little less than the last one did.

At least one of the giants, as we will see, is deliberately giving capable models away for free, setting the price of good-enough intelligence at zero on purpose, the way Insull drove the price of a kilowatt-hour toward the floor. Put those forces together and an equilibrium emerges, in which base intelligence becomes cheap, abundant, metered, and interchangeable. A commodity. It is the exact fate of bandwidth and electricity and transport before it.

Consider what that means for the company that climbed all the way to the top of the stack to escape the trap. It climbed from the grid to the platform to the model, following the value upward the way GE did, doing everything a smart infrastructure owner should do, and it arrives at the summit to discover that the summit is commoditizing underneath its feet too. There is no rung you can reach and rest on. GE owned the appliances, and then the appliances commoditized, and the value went to brands and retailers and designers above even them.

The stack does not have a top where the migration politely stops. It has one more layer, always, and the value is always climbing toward it. That is the real question the rest of this book exists to answer, the question that the hyperscalers, for all their power, cannot escape any more than the railroads could. If the grid commoditizes, and the platform commoditizes, and the model itself commoditizes, then where, in the age of artificial intelligence, does the durable value finally go?

The Sword, in a New Form

Before we answer that one, there is one more piece of the internet story that belongs here, because it shows the Indispensability Trap operating in a form we have not quite seen before, adapted to a world of software.

In the earlier chapters, the essential infrastructure was physical and singular. There was one electrical grid, one telephone network, one set of rails into your town, and because the thing was a physical bottleneck that could not be routed around, society reached for the heaviest tool it had, rate regulation, and told the owners what they could charge. But the dominant companies of the internet age were not physical bottlenecks. They were software, and software is substitutable in a way that a power line is not. There were other operating systems besides Windows. There were, in principle, other search engines besides Google. And because the dominance was real but the thing was substitutable, society did not rate-regulate these companies as utilities. It disciplined them with a different tool, the antitrust suit, and the credible, permanent threat of one.

The pattern was set a full generation before the dot-com boom, with IBM. In 1969 the Justice Department accused IBM of monopolizing the market for general-purpose computers, and the case ground on for almost thirteen years before it was dropped in 1982 as "without merit," a marathon one exhausted lawyer famously called the antitrust division's Vietnam.

IBM was never broken up and never lost in court. But the case did something subtler and more lasting, it made the most powerful company in computing cautious, taught it to move with a lawyer at its elbow, and that caution shaped the decision that gave away the next era. When IBM built its personal computer, it chose not to demand exclusive ownership of the operating system, and let a small supplier named Microsoft keep the rights to license DOS to everyone else. A more aggressive IBM, one not shadowed by a decade of antitrust litigation, might never have handed its successor the keys. The giant's caution is what let the next giant grow.

And the next giant grew into exactly the same position. By the late 1990s Windows was so dominant, ran on so overwhelming a share of the world's computers, that Microsoft had become genuinely indispensable to the personal computer, and it used that position in ways the government decided were illegal, most famously by crushing the Netscape web browser to protect its franchise. It became the target of the monopoly police in its turn.

The Justice Department sued Microsoft in 1998, and in 2000 a federal judge ruled Microsoft an unlawful monopolist and ordered the company broken in two. That breakup was overturned on appeal, and the case was ultimately settled with behavioral remedies rather than dismemberment, but the finding of monopoly stood, and the effect on Microsoft was profound and lasting. For a decade afterward, Microsoft operated with a lawyer always in the room, cautious about using its power too nakedly, its every competitive move shadowed by the memory of the trial. And that caution, many have argued, is precisely what created the opening through which its successors climbed. A Microsoft unafraid of the government might well have strangled Google in its cradle, the way it had strangled Netscape. A Microsoft on a leash let Google grow.

And then, in the fullness of time, the sword swung toward Google itself. In August of 2024 a federal judge ruled that Google, too, was an illegal monopolist, in its case over internet search, the first successful monopoly case against a technology giant since Microsoft. And in a detail that ought to make every executive in the industry pay attention, the Justice Department built its case against Google using the Microsoft case as its explicit legal roadmap, and the judge's opinion cited the Microsoft precedent directly.

The instrument that disciplined the dominant company of one generation was picked up, barely changed, and pointed at the dominant company of the next. IBM, then Microsoft, then Google. Three turns of the same wheel. The trap does not go away because the infrastructure became software. It changes its enforcement mechanism, from the rate commission to the antitrust court, but the underlying logic holds. Become sufficiently indispensable, and the state will find a tool to cap you, whether that tool is a regulated rate of return or a monopolization suit. And the sword, having passed from IBM to Microsoft to Google, is even now being sized up for whoever becomes indispensable in artificial intelligence.

The Roads Were Already Built

Which brings us, finally, to the layer that was quietly assembling itself underneath this whole story, and to the reason this chapter carries the title it does.

While the fiber was crashing and the cloud was rising and the antitrust suits were flying, a large and loosely coordinated community of engineers spent about fifteen years solving an unglamorous set of problems. Solving the problem of how to package a piece of software so it runs the same way everywhere, which produced, around 2013, the technology called containers, and the company, Docker, that popularized it.

The next problem was how to coordinate thousands of those containers across thousands of machines without a human managing each one. Google, drawing on a system it had built internally to run its own vast operations, released its solution as open source to the world in 2014 as Kubernetes. This begat the problem of how to give ordinary developers a smooth path onto all this machinery without requiring each of them to become an expert in it, which produced the disciplines that came to be called DevOps and platform engineering.

These are not household names. Almost no one outside the industry has heard of them. But together they constitute something momentous, an operating system for running software at planetary scale, a set of standardized rules and tools for how enormous volumes of computation move across a distributed world without constant collision.

I have spent a good part of my career documenting the community that built this layer, and I have come to think of it in a particular way, which I want to offer as the bridge from this chapter into the rest of the book.

Imagine the cloud-native (as this system of containers and Kubernetes came to be called) era as the construction of an interstate highway system. For fifteen years, a large community poured the roadbed, set the standards for lanes and signage, built the interchanges, and worked out the rules that let enormous volumes of traffic move across a continent without constant collision. It was slow, unglamorous, foundational work. The people doing it were not building toward any single destination. They were building the surface on which future movement would happen, whatever that movement turned out to be. And by the time they were largely done, the highway had become invisible, in the way that infrastructure does when it works.

Artificial intelligence is the first generation of software capable of driving on that highway autonomously. The large language models, the AI agent, the whole apparatus of machine intelligence that the next twelve chapters are about, did not arrive on virgin ground and have to build its own foundations. It arrived to find the roads already poured, already engineered for scale, already governed by a working set of rules about identity and security and traffic flow, all of it built by people who had never heard of a transformer model and were solving a completely different problem. The scheduling that assigns AI workloads to machines, the identity systems that decide what an AI agent is allowed to touch, the observability that lets you see what it is doing, the resilience that keeps it running when a machine fails, all of this was built by the cloud-native community over the fifteen years before AI needed it. The AI model is the vehicle. Cloud native is the road. And the roads were already built.

Look at what that means through the lens of this entire book. The AI grid that everyone is watching in 2026, the $200 billion pouring into Richland Parish and the nine gigawatts planned for the Utah desert, the whole spectacle of the physical build-out, is riding on top of fifteen years of operational engineering that got poured while almost no one was paying attention. It is value migration made literal, a stack of abstractions in which each layer hides the complexity of the one beneath it so the next can be built. Hardware at the bottom. Then virtualization hiding the hardware. Then the cloud hiding the infrastructure. Then containers hiding the packaging. Then Kubernetes hiding the distributed coordination. Then platform engineering and DevOps hiding Kubernetes. And now artificial intelligence, the newest layer, resting on all of it, hiding the machinery of reasoning itself. Each layer was somebody's frontier once, somebody's expensive, dangerous, cutting-edge achievement, and each one has become the boring, assumed, invisible foundation for the layer above.

That is the deepest form of the lesson. The value does not just migrate sideways, from the infrastructure company to the company that builds on top. It migrates upward, layer by layer, each generation's hard-won frontier becoming the next generation's free assumption. The fiber that bankrupted Global Crossing became the free backbone under Google. The virtualization that VMware fought to invent became the invisible foundation under Amazon. The Kubernetes that Google built to run its own empire became the free operating system under a thousand AI startups. And the artificial intelligence being built at such staggering cost today, the intelligence the whole world is transfixed by, is on its way to becoming exactly the same thing, the expensive, dangerous frontier of this moment, and the cheap, invisible, assumed foundation of the moment after that. It is precisely where the rest of this book is headed.

This story started, for me, in a conference room in Houston, watching a doomed company try to trade a commodity that was collapsing in value faster than they could sell it. I built a company myself on the layer below the one where the money would eventually be, then reached for the layer above just before the tool that would have made it possible was invented, and got washed under for my timing.

I have spent the years since watching an entire industry pour the roads that artificial intelligence would later drive on, without anyone quite realizing that was what they were doing. If there is one thing I know in my bones rather than from a book, it is this: The people building the most visible, most expensive, most celebrated layer of a technology are almost never the ones who end up with the lasting fortune. The fortune goes up the stack, to the ones who treat today's miracle as tomorrow's plumbing. It happened to the railroads and the telephone and the electrical grid, and it happened to the fiber I bought and tried to sell in Houston, and the only real question left in this book is whether it is about to happen again, on the grandest scale yet, to artificial intelligence.

Part Three: The Intelligence Grid

Whether owning the full AI stack finally lets an infrastructure owner keep what it built.

Chapter 7

The Fork in the Road

> The strategic inflection point is the time to wake up and listen.
>
> —Andrew S. Grove, *Only the Paranoid Survive*, 1996

Goldman Sachs put a number on the build-out: roughly $7.6 trillion in cumulative AI infrastructure spending through 2031. Hold that figure loosely. It moves fast, and it undercounts the second tier of builders now crowding into the same build-out. But even loosely held, it establishes the scale of what closes out Part II. Railroads, telephone lines, electrical grids, and the internet's fiber and server infrastructure were each, in their moment, the largest capital commitment their economy had ever made to a single technology. AI is now that commitment for this one.

The question asked of every prior build-out is the question asked again here: When the infrastructure becomes indispensable, who keeps the money?

The last chapter ended on a fork. The internet and cloud era showed the grid commoditizing, with bandwidth, storage, and compute all becoming cheaper and more abundant, exactly as Jevons paradox would predict, while the fortunes accrued not to the companies that laid fiber or ran data centers but to the ones that built on top of them, treating the infrastructure as a free input. Google didn't own the pipes. Amazon didn't own the backbone. They built where the commodity was cheapest and captured the value the infrastructure owners couldn't hold onto.

AI appeared, at first glance, to offer the infrastructure owners a second chance. This time the hyperscalers—Google, Microsoft, Amazon, and Meta—are not commodity providers waiting to be disintermediated. They're the builders. They own the data centers, the custom silicon, the power contracts, and, increasingly, the frontier models themselves. The bet embedded in a trillion dollars of capex is that owning the full stack, top to bottom, finally allows an infrastructure owner keep the wealth it built.

Part III exists to test that bet against the same four engines that have explained every prior build-out: the Indispensability Trap, Jevons paradox, the Perez installation-to-deployment cycle, and the distinction between financial and production capital. The test starts with a harder version of the fork than the last chapter needed to resolve, because AI's fork isn't binary. It's got three prongs, and they're collapsing onto the same handful of companies at once.

The Wider Grid

The first pressure is the one already previewed: The physical and computational grid underneath AI is getting more crowded, not less. It isn't just Google, Microsoft, Amazon, and Meta. It's neoclouds like CoreWeave. It's Oracle, a company that spent two decades as a punchline in cloud circles and is now one of the more aggressive capacity builders in the industry. It's colocation providers, sovereign compute initiatives, and, as of February 2026, SpaceX, which absorbed Elon Musk's xAI in an all-stock merger and immediately became a GPU landlord to two of its own most direct competitors.

The arrangement is stranger than anything the railroad or telephone era produced. Google, arguably the single largest owner of AI compute on the planet, on the strength of its custom Tensor Processing Unit (TPU) silicon, is paying SpaceX roughly $920 million a month for access to some 110,000 Nvidia GPUs housed at xAI's data centers, a deal that runs from October 2026 through June 2029 and totals somewhere north of $30 billion. Anthropic, in a similar arrangement announced weeks earlier, is paying SpaceX $1.25 billion a month to rent the entirety of xAI's Colossus 1 facility, with access to up to 220,000 Nvidia GPUs. SpaceX's IPO paperwork framed the deal as running through 2029; Musk has since publicly described it as a six-month lease instead, and both this arrangement and the Google deal are cancelable by either party with ninety days' notice. Those GPUs were presumably built for SpaceX's Grok model, but Grok has fallen well behind the frontier, and SpaceX appears to be hedging by renting out capacity rather than betting it all on a model that isn't winning.

Why rent out the very GPUs Grok needs? Brad Feld, a venture capitalist I've learned from over the years, calls it the rule of three: In any tech market, if you're not in the top three, get out. First place takes the lion's share of the money the market attracts, second place earns a healthy return, third place a decent one, and after that valuations sink and investors are lucky to get their principal back. Grok is not in AI's top three. Some argue AI is different enough that the rule won't hold, but it goes a long way toward explaining why Musk and SpaceX's leadership are quietly repositioning from model maker to GPU landlord.

Meanwhile, Google is selling TPU capacity to Anthropic in the other direction, a deal reported at $40 billion for access to up to a million Ironwood chips over five years, while its own DeepMind researchers wait weeks for training-run allocation because paying external customers gets priority.

Google is simultaneously the grid's biggest owner, a tenant on someone else's grid, and a landlord to its own rival. No single label, whether infrastructure owner, platform builder, or model maker, holds still long enough to describe what any of these companies actually are in 2026. That instability is itself evidence for this book's thesis, not a complication of it. It's what a commodity in the middle of a Perez installation frenzy looks like from the inside: Capital moving faster than roles can settle, and every player hedging against being the one left holding depreciating infrastructure when the frenzy ends.

The Free Model

The second pressure is newer than anything the railroad, telephone, or electrical chapters had to reckon with, and it's the one that keeps this book from collapsing into a simple "climb the stack" argument. In those earlier build-outs, there was always a layer above the commodity that stayed defensible: the freight brokerage, the long-distance network, the appliance built to run on standardized current. AI's added twist is that the layer everyone assumed would be defensible, the frontier model itself, is commoditizing too, and faster than most forecasts anticipated.

By mid-2026, models trained and released under fully permissive licenses were landing within fractions of a benchmark point of the best closed, proprietary systems money could buy. Frontier convergence, once treated as a multiyear gap, had narrowed to single digits. Every major lab, American, European, and Chinese, was arriving at the same mixture-of-experts architecture independently, the way competing railroads all converged on standard gauge once the alternative got expensive enough. Diminishing returns on brute-force data and compute scaling pushed labs toward efficiency gains that, once published, were trivially copyable. And Mistral, the French lab that spent two years building a business on restrictive, revenue-gated licensing, gave up and released its flagship line under full Apache 2.0 in December 2025, not because it wanted to, but because holding the line was costing it the ecosystem.

If the model is also a commodity, there's no safe rung on the ladder. That's Chapter 10's argument in full, but I bring it up here because it reframes everything the rest of Part III examines. This isn't a story about infrastructure owners losing to platform builders the way railroads lost to Vanderbilt's arbitrage or telephone equipment lost to Vail's network effects. It's a story where the commoditization wave doesn't stop at any layer. It keeps climbing, washing over successive layers of wealth creation. The only question left is whether anything above the model itself is durable enough to hold value at all.

Sovereign Ground

The third pressure is the one that none of the earlier build-outs quite faced, though the railroad chapter's land grants and the telephone chapter's regulatory bargains are its distant ancestors. As AI infrastructure concentrated in the hands of a handful of American companies, nations started treating that concentration as a strategic vulnerability rather than a market outcome. The United Kingdom's technology secretary stood up in 2026 and said the quiet part out loud: Seventy percent of global AI compute sits with five companies, and no country that wants to keep its independence can accept that arrangement indefinitely. The European Union backed the sentiment with tens of billions of euros for state-anchored AI gigafactories explicitly designed to be public infrastructure rather than hyperscaler property. Canada, moving in the same window, announced its own multibillion dollar sovereign compute strategy. Roughly ninety countries signed a declaration committing to AI sovereignty as a matter of formal policy.

This is the Indispensability Trap's access-guarantee mechanism showing up a full technological generation ahead of where the railroad and telephone build-outs found it. Rate regulation arrived only after decades of freight abuse. The Bell System's universal service bargain took the threat of a hostile Justice Department to negotiate. Sovereign AI didn't wait for abuse; it's being built preemptively, before the indispensability is fully installed, because the nations building it watched what the earlier build-outs did to someone else's infrastructure and they don't intend to repeat the position.

And the access-guarantee mechanism has a second face here that the earlier build-outs never had to reckon with in real time: export control. In June 2026, Anthropic released its Fable and Mythos models, then had to suspend access to both within days to comply with US Department of Commerce controls. The restrictions were lifted at the end of the month, restoring access on July 1. The controls came and went in three weeks. But the episode is a preview of something structural: The model layer isn't just exposed to market commoditization from below. It's exposed to state authority from above, and that authority can act faster than any regulatory body examined so far. A rate case takes years. An antitrust suit takes a decade. An export control designation takes a signature.

The Map

Three pressures. One set of companies. Google, Microsoft, Amazon, Meta, and Oracle sit on one side of a rough divide. They own physical infrastructure, and the Indispensability Trap's classic mechanism (get essential enough and society caps your return) applies to them most directly. OpenAI and Anthropic sit on the other side. They rent the grid rather than own it, and the mechanism that catches them is model-layer commoditization and, as we will see in Chapter 13, state leverage over what they're allowed to ship. xAI, folded into SpaceX, straddles both, infrastructure owner and model maker in one balance sheet, with Musk himself signaling that he intends to keep most of the Colossus 2 facility walled off from the largest leasing arrangements funding the rest.

Eight companies, three pressures, and, per the pattern, no clean answer yet about where the money lands. The railroad chapter took Vanderbilt's arbitrage instinct to find the answer. The telephone chapter took Vail's regulatory bargain. The electrification chapter took Insull's overreach and collapse. The internet chapter took an operator's firsthand view of the commodity trap closing in real time. Part III doesn't have a single protagonist to follow through the AI build-out. The field is too wide and too recent for that. What it has instead is the map itself: Seven companies making seven different bets about which layer will hold value once the grid, the platform, and the model have all been commoditized out from under them. If history is any guide, and if a career in technology has taught me anything, the real winner in AI may come from an industry or class of product we haven't seen yet, and it will be built atop the commoditizing layers below.

The next chapters open that map wall by wall. Chapter 8 stays with the landlord-tenant-rival tangle at the infrastructure layer. Chapter 9 goes one layer down, to the silicon substrate underneath the cloud contracts and model equity. Chapter 10 goes to the model layer and the open-weight wave that's commoditizing it. Chapter 11 follows that wave to its most state-directed instance, in China. Chapter 12 follows the nationalist response it's provoking everywhere else. Chapter 13 returns to the export control precedent, and what it means that the sword can now fall on a model instead of a merger. Chapter 14 looks at a company who does appear to have it all, is not one of the MAGOS 8 companies mentioned and is in fact bigger than any of them by market cap. But is the trap already closing in on it? Chapter 15 closes Part III by asking, with all three pressures on the table at once, what's actually left for these eight companies, (let's call them the MAGOS 8) to fight over, and where each of them is betting the fight will be won.

Chapter 8

Landlord, Tenant, Rival

> Financial genius is before the fall.
>
> —John Kenneth Galbraith, *A Short History of Financial Euphoria*, 1990

On January 29, 2026, Microsoft's stock fell roughly 10 percent in a single trading session, wiping out about $357 billion in market value in one day, the company's worst single-day dollar loss in its history and its steepest quarterly decline since the 2008 financial crisis.

The trigger wasn't disappointing revenue. Its cloud service Azure had grown 39 percent, ahead of forecasts. Its cash cow businesses in Windows and Office 365 were on target. All of the other revenue producing arms of the Redmond giant were in line. The trigger was what Microsoft was spending. Capital expenditure had hit $37.5 billion in the quarter, up 66 percent from a year earlier. Investors had started asking a question that would define the rest of 2026 for every company in this chapter: Was this spending building something that pays for itself or was it just a very large bet that hadn't been called yet?

Buried inside that anxiety was a relationship even stranger than the number suggests, the details of which had been rewritten twice in the preceding six months. Microsoft had backed OpenAI since 2019 under an arrangement built around a single, undefined trigger. If OpenAI's board ever declared it had achieved artificial general intelligence, Microsoft's exclusive license to OpenAI's technology would end. Nobody had defined AGI in the contract. OpenAI could, in principle, make that declaration unilaterally and walk away from its biggest backer.

In October 2025 the two companies rewrote the deal. OpenAI converted into a public benefit corporation, a structure that let it raise capital more like an ordinary company while a nonprofit foundation retained oversight, and Microsoft came out the other side holding roughly 27 percent of the new entity, valued near $135 billion, down from a nominal 32.5 percent before the round diluted it. The dilution nonetheless produced a $5.9 billion accounting gain for Microsoft, because OpenAI's implied value had grown faster than Microsoft's percentage of it had shrunk. The AGI clause survived that round in modified form, routed through an independent verification panel instead of OpenAI's say-so.

It didn't survive the next round. Six months later, on April 27, 2026, the two companies rewrote the deal again. The AGI clause was gone entirely. Microsoft's license to OpenAI's models now ran on a fixed calendar, through 2032, regardless of what OpenAI later claimed about its own capabilities. In exchange Microsoft gave up its exclusive right to host and resell OpenAI's models. OpenAI became free to ship its products through any cloud provider it chose, including Microsoft's hyperscaler rivals. The timing was not subtle: AWS announced OpenAI's frontier models live on Bedrock the very next day, the direct payoff of a $50 billion investment Amazon had made in OpenAI two months earlier.

Around the same set of announcements, reporting surfaced that OpenAI had missed several of its own internal revenue targets, that ChatGPT's weekly active users had flattened well short of a billion, and that OpenAI's chief financial officer had told colleagues she worried the company might not be able to fund its future computing contracts if revenue didn't accelerate. Sam Altman and his CFO jointly called the report "ridiculous" and said they were "totally aligned on buying as much compute as we can." It was a denial that answered a question nobody had asked and left the one that mattered, whether OpenAI could actually afford what it had committed to buy, sitting exactly where it had been.

Layer the product decisions on top of the contract rewrites and the picture gets stranger still. In March 2026 Microsoft launched Copilot Cowork, an autonomous agent product built to handle a quarter's worth of a person's meetings and email without supervision. Its most demanding reasoning work doesn't run on OpenAI's models. It runs on Claude, built by Anthropic, OpenAI's most direct competitor. Charles Lamanna, Microsoft's president of business apps, described the move as a deliberate hedge. The company had watched its

OpenAI-only strategy start to look fragile and wanted to make sure its own customers weren't locked into a single lab's fortunes. Weeks before that, on April 2, Microsoft had already begun shipping its own family of in-house models (MAI-Voice-1, MAI-Transcribe-1, MAI-Image-2) built for the explicit purpose of reducing its dependence on the company it had just spent six months renegotiating a $135 billion stake in. By the middle of 2026, Microsoft was, by its own design, selling frontier AI capability sourced from three rival labs inside a single product line: OpenAI, in which it owns a large and newly nonexclusive stake; Anthropic, from which it simply buys; and itself.

At Microsoft's Build developer conference that same season, the company made what one industry analyst called the architectural version of the quiet part said out loud: It did not intend to be merely the reseller of somebody else's intelligence. That is an unusual thing for the largest shareholder of the intelligence in question to feel the need to say aloud. It is also the clearest evidence so far that owning a stake in the model layer does not settle the question of where value ends up. Microsoft has the equity, the cloud contract, and the distribution, and it still doesn't trust any of the three positions enough to stop hedging against all of them simultaneously.

The Federal Trade Commission noticed a version of the same tension from a different angle. An investigation opened in late 2024 into whether the Microsoft-OpenAI structure was built specifically to avoid the merger review a straightforward acquisition would have triggered—an investment large enough to function as effective control while dressed in the language of a minority stake. That investigation has continued across changes in FTC leadership. That kind of survival across an administration change suggests it isn't a partisan exercise. It's a standing question about whether this industry's closest partnerships are quietly doing the legal work of mergers without asking anyone's permission for one. This concept resurfaces later in this book at exactly the moment the antitrust sword comes looking for its next target.

The Same Playbook, Twice

Microsoft's tangle looks unique until you notice that Amazon ran an almost identical version of it twice within the same ninety days, with the two companies that least want to be treated identically.

In February 2026 Amazon invested $50 billion in OpenAI, and OpenAI committed to two gigawatts of Trainium silicon, positioning AWS to become a distributor of OpenAI's enterprise platform once Microsoft's exclusivity lapsed. Two months later, on April 20, Amazon announced what was, in every structural respect, the same deal with OpenAI's chief rival. Amazon would put up to $25 billion in new investment in Anthropic, on top of the $8 billion Amazon had already put in since 2023. Anthropic would commit to spend more than $100 billion on AWS technology over the following decade and secure up to five gigawatts of Trainium capacity to train and run Claude. One trade outlet's headline captured the moment precisely: Amazon doubles down on Anthropic, mirroring its OpenAI cloud deal. Andy Jassy signed both announcements within a single fiscal quarter.

Amazon is not a neutral landlord in either relationship. It ships its own Nova family of foundation models directly inside Bedrock, the same marketplace where more than 100,000 customers already run Claude, competing for the same enterprise workloads as the labs Amazon is simultaneously bankrolling.

Google occupies the mirror image of this position with Anthropic: a major investor and TPU supplier and the maker of Gemini, a direct competitor to the model it partly finances. By the middle of 2026, every hyperscaler with a frontier ambition of its own—Microsoft, Amazon, and Google—had become an investor in at least one rival lab, a cloud supplier to that same lab, and a competitor to it, all at once, and all three had reached that position through their own independent negotiations rather than by copying each other. That convergence is its own kind of evidence. When three companies with different histories, different cultures, and different boardrooms arrive at the identical hedge within the same year, the hedge isn't a strategic insight unique to any one of them. It's what the shape of the market currently requires of anyone large enough to be in this game.

The Circle

Underneath the blurred roles sits a financial structure that by 2026 had earned its own name in the trade press: circular AI financing. The mechanism is simple enough to sketch. A chipmaker or cloud provider invests cash or equity in an AI lab. The lab spends that same money, or a close approximation of it, buying the investor's own chips or cloud capacity. Both sides record the transaction as revenue, backlog, or committed demand. The dollars complete a loop and arrive close to where they started, having touched very little of the world outside the small group of companies running the circuit.

By 2026 analysts had priced the scale of this at more than $800 billion, spread across a dense web of overlapping deals well beyond the Microsoft and Amazon relationships already described. Nvidia sits at the center of the largest single strand with a planned $100 billion investment in OpenAI, structured so that OpenAI would use much of the proceeds to buy Nvidia chips. In February 2026 a report that this particular deal had stalled was enough to send Nvidia, OpenAI, and Oracle all wobbling within the same news cycle, even though none had actually lost a contract. Oracle went so far as to issue a public statement insisting the stalled talks had zero impact on its own relationship with OpenAI, an unforced denial that did more to confirm the market's fear than to calm it. Altman, for his part, took to social media to call the panic nonsense and describe Nvidia as OpenAI's most important partner in both training and inference.

Nvidia's exposure runs well past OpenAI. It has taken stakes in xAI and in Mistral, both of which remain dependent on Nvidia's chips to operate. It has backed the funding rounds of neocloud providers including Nscale and Nebius, taken a roughly 7 percent stake in CoreWeave, and separately agreed to buy $6.3 billion of cloud services from that same company, a provider whose entire business consists of renting out Nvidia hardware.

CoreWeave itself sits inside another loop. It handed OpenAI $350 million in equity ahead of its own IPO, and OpenAI's cloud commitments back to CoreWeave have since grown past $22 billion. AMD, for its part, handed OpenAI warrants to acquire AMD stock at a nominal price in exchange for OpenAI committing its compute build-out to AMD's chips. One financial writer noted that the arrangement predictably increases AMD's market value the moment it is announced, meaning OpenAI effectively profits from making an announcement that benefits its counterparty before a single chip ships.

Oracle, meanwhile, signed a $300 billion, multiyear infrastructure commitment with OpenAI to build out Stargate capacity, a deal so large relative to Oracle's balance sheet that doubts about its ability to deliver sent Oracle's stock down roughly 30 percent in a single quarter.

None of these deals is secret. Each one is disclosed, in some form, in a public filing, an earnings call, or a press release. What's harder to see from inside any single disclosure is the aggregate, the same relatively small group of companies appearing on multiple sides of multiple deals, each one booking the same underlying dollars as new revenue, new backlog, or new demand. A dollar that starts in Nvidia's treasury becomes an investment in a lab, then a lease payment to a cloud provider, then revenue that provider reports to its own shareholders. By the time it settles anywhere, it's been counted at least twice. Multiply that by $800 billion of overlapping structures and the honest answer to how much of this represents demand from outside the circle, by real people and companies paying for AI services because they need them rather than the circle counting itself, becomes very hard to isolate. If this doesn't sound like the table at that meeting in Houston with Enron, I don't know what does.

The Guarantee

Broadcom's version of the circle goes one step further than any of the others, and the extra step is worth walking through slowly, because it is the difference between participating in a boom and personally underwriting it.

In 2026 Broadcom joined Apollo and Blackstone to build what the three companies called the AI XPV Platform, a special-purpose vehicle whose first transaction raised $35 billion to buy Google TPUs and lease them to Anthropic. Structurally, this is an ordinary equipment financing arrangement: A fund buys hardware, leases it to the company that needs it, and repays its own debt out of the lease payments. Broadcom's chief executive, Hock Tan, called it the first of many such transactions. He said the platform was intended to eventually finance more than 20 gigawatts of AI infrastructure through 2028, a figure that implies something on the order of $700 billion in eventual chip procurement if the ambition is realized in full.

What makes this particular structure different is that Broadcom personally guaranteed the senior layers of the debt. The $35 billion raise was split across three tranches: roughly $6 billion in senior notes sold to banks, another roughly $24 billion in a second senior tranche, and a final $4.5 billion junior tranche that carried no Broadcom backing at all. On the two senior tranches, Broadcom provided what's known in the market as a deficiency guarantee, a promise that if Anthropic ever falls behind on its lease payments and the underlying chips can't be resold for enough to cover the shortfall, Broadcom's balance sheet makes up the difference. That guarantee is precisely what let the senior notes price at a coupon around 5.75 percent, a rate that reflects Broadcom's strong credit rather than Anthropic's considerably riskier stand-alone profile. The unguaranteed junior tranche, exposed to the same underlying risk without Broadcom's backstop, priced at 8.5 percent, nearly three full points higher. The gap between those two numbers is the market's honest estimate of how much of this deal's safety exists only because Broadcom is standing behind it.

Standard & Poor's responded within days by downgrading Broadcom's credit rating, citing the leverage the guarantee had quietly added to Broadcom's books. Broadcom's balance sheet, carrying roughly $65 billion in debt against under $20 billion in cash, has considerably less room to absorb a disappointment than Nvidia's does. A single guarantee on a single customer's chip lease had become a rating agency's concern about Broadcom itself.

The same pattern showed up, in rawer form, on the OpenAI side of Broadcom's business. Broadcom has long operated under a standing internal rule for any customer seeking financing for a custom chip project: The customer puts up one dollar of its own money for every dollar Broadcom contributes, a standard piece of lender discipline meant to keep Broadcom from ending up more exposed to a customer's fortunes than the customer itself is.

For OpenAI's custom chip program, internally known as Project Nexus and aimed at more than 10 gigawatts of compute before 2030, Broadcom dropped that rule. It ended up contributing more capital up front than OpenAI did, a break from what one report called Broadcom's long-held hardline requirement. The immediate trigger, reportedly, was Microsoft's refusal to provide the purchase guarantees that would have made the financing bankable on ordinary terms, leaving Broadcom to either walk away from a customer it very much wanted or absorb the risk itself. It chose to absorb it.

Line the two arrangements up and Broadcom is not merely selling chips into a boom, the way Nvidia primarily is. It has become, in its own accounting and now in a rating agency's judgment, one of the underwriters of its customers' ability to keep buying. That's a materially different position from investing in a customer, because an investment can be marked down without threatening the lender's credit rating. A guarantee cannot. Broadcom didn't get pulled into this circle by accident. It built one of the load-bearing walls, and it built it out of its own balance sheet.

The Other Side of the Argument

None of this proves a bubble is forming, and the strongest voices making the opposite case deserve to be heard on their own terms rather than waved off.

One line of defense focuses on where the money actually originates. Several analysts have pointed out that the great majority of these circular arrangements involve two large, disclosure-bound private or public balance sheets trading claims with each other, not a company selling shares to retail investors who can't see the whole picture. One widely read commentary on the subject proposed a simple test borrowed from prior bubbles: Watch for a wave of AI companies going public, watch for firms making acquisitions using their own inflated stock as currency, and watch for the biggest tech companies to stop buying back their own shares, because all three showed up clearly in 1999 and 2000.

Through the first half of 2026, that argument held. Then, in a single week in June, two of the three signals fired.

SpaceX priced its IPO on June 12 at a $2.1 trillion valuation, raising $86.2 billion in what became the largest offering in market history. Four trading days later, on June 16, with its own share price up more than 40 percent from the offer, SpaceX exercised a prenegotiated option to acquire the AI coding company Cursor for $60 billion entirely in stock, the largest acquisition of a venture-backed startup on record. Both the IPO and the stock-as-currency M&A that the counterargument's test had said were absent turned out to be a few weeks away from happening at unprecedented scale. That does not by itself prove the market is in a bubble. It does mean the defense that rested on their absence needs to be re-evaluated on the terms the defense itself proposed.

A second line of defense looks at scale relative to the companies involved. UBS's analysis of the Nvidia-OpenAI arrangement put it at somewhere around 13 percent of Nvidia's projected 2026 revenue. A meaningful number, but not one that suggests Nvidia's entire business rests on a single circular transaction. The AMD warrant deal, on close inspection, involves OpenAI receiving equity from AMD rather than AMD selling shares into the market, which is close to the opposite of the pattern regulators usually worry about in a genuine bubble. And the underlying demand, on this view, is not manufactured by the financing at all.

Anthropic's annualized revenue reportedly climbed from roughly $1 billion to $5 billion in under six months in 2025, then past $30 billion by April 2026. Google disclosed that its monthly token processing grew roughly sevenfold in a year, from 480 trillion tokens in May 2025 to 3.2 quadrillion by May 2026.

If real usage is climbing that quickly, the argument goes, the financing isn't inflating a phantom market. It's a fast way of building capacity that a slower, more conventional fundraising process couldn't keep pace with. Really no different in kind from railroads raising capital ahead of confirmed freight volume or utilities building generating capacity ahead of confirmed household demand.

Both of those defenses are serious, and both rest on the same underlying assumption, that the demand curve now visible in revenue statements keeps climbing at close to its current rate. OpenAI, by widely cited reporting, is on track to lose in the neighborhood of $14 billion in 2026, nearly three times its prior year's loss, against a projection of $100 billion in revenue by 2029.

This is the same company whose CFO privately worried about funding its own compute contracts even as its CEO publicly dismissed the worry as nonsense. Every arrangement described in this chapter is, at bottom, a bet that the distance between today's losses and future revenue closes because paying customers outside the circle keep showing up in growing numbers, not because the next round of vendor financing keeps the appearance of growth alive for one more quarter.

If the demand holds, this becomes exactly what its defenders describe, an efficient, if unusually tangled, way to prefund a capital-intensive build-out, the same kind railroads and electrification required and ultimately survived. If it doesn't, the guarantees, the warrants, and the leased gigawatts stop being separate claims on separate companies and become, all at once, claims on the same shrinking pool of real revenue.

This book has borrowed Carlota Perez's framework since the beginning to describe how every major technological build-out moves through an installation phase before either correcting hard or settling into deployment. In the installation phase, financial capital races ahead of any confirmed use for what it is funding. In the deployment phase, production capital, the actual roads and wires and plants, catches up to justify the spending that came before it. Every build-out has had its own installation frenzy, the railroad mileage laid well ahead of the freight needed to fill it, the parallel long-distance telephone lines strung by competing companies before either could turn a profit, the utility holding companies stacked one atop another until Samuel Insull's empire collapsed under leverage nobody had bothered to trace to its source.

Circular AI financing reads as this generation's version of the same signal. It is financial capital, in the most literal sense, since the dollars in question are mostly commitments, guarantees, and equity stakes rather than completed data centers or shipped chips. They are racing around a loop fast enough that distinguishing it from genuine production capital requires waiting to see what survives the next downturn.

The Perez framework doesn't say a frenzy like this is necessarily a mistake. Every build-out needed one to get built at all. What it does say is that a frenzy always ends in one of two ways: a hard correction that strands the most leveraged players, the way it stranded Global Crossing and WorldCom, or a slower unwind that leaves the underlying infrastructure standing even as the paper claims on top of it get sorted out. Which of those two paths AI's circular financing takes is not yet knowable. That it is currently indistinguishable from the early stage of either path is the whole point.

Where This Has Happened Before

None of this should feel unfamiliar. In the late 1990s telecom equipment makers financed the very carriers buying their gear, propping up a demand curve that looked organic for years until it wasn't. Some carriers went further, swapping fiber capacity with each other and booking the trades as sales even when both sides were effectively buying back what they had already sold just to keep growth numbers pointed in the right direction. Congressional investigators later examined exactly this tactic at Qwest and Global Crossing. When real usage failed to catch up with the capacity everyone had financed into existence, the most leveraged carriers went first, and the fiber they had built sat largely dark for years afterward while the value that survived climbed to companies, Google and Amazon among them, that had never owned a mile of it.

By 2026 that comparison had stopped being a historian's footnote and started coming from people whose job is to price risk rather than narrate it after the fact. Michael Burry, the hedge fund manager who built his reputation shorting housing in 2008, disclosed in December 2025 a short position against Nvidia, explicitly comparing Nvidia's strategy of financing its own customers' purchases to Enron's practices in the years before its 2001 collapse. Bloomberg's reporting on the circular deals drew the fiber glut comparison directly, naming Qwest and Global Crossing without hedging.

I sat across a conference table from the company that comparison invokes, in a high floor overlooking downtown Houston in the spring of 2000 and watched a bandwidth arbitrage pitch that made no sense on its own economics. Buy your own capacity at your own price, pass it around the table, sell it back at half. My business partner's read on the room afterward, that the people across the table were either the smartest he'd ever met or the biggest crooks, has stayed with me for a quarter century, because it turned out both things could be true of the same room at the same time. What I didn't fully appreciate until years later is that the room wasn't unique to Enron. It's what a frenzy looks like from the inside, at any scale, in any decade. Revenue generated by financing your own customer's purchase of your own product looks exactly like organic demand for as long as nobody actually needs the money to come from somewhere else. It stops looking like organic demand the moment growth so much as pauses, and by then the companies most leveraged into the circle rarely get much warning.

The Bet Nobody Can Call Yet

The distinction between financial capital chasing a return on a technology's promise and production capital that has actually become the roads, wires, plants, and data centers the technology runs on, is the only lens that ultimately resolves what's happening in this chapter. But it can't be applied yet, because the resolution only becomes visible after the correction that tests it. The railroads absorbed a financing mania and left behind track that outlived the speculators who financed it. The fiber glut absorbed a financing mania and left behind dark cable that outlived Global Crossing. Whether AI's $800 billion in circular commitments converts into data centers, chips, and models that keep working long after the financing arrangements that built them have been renegotiated, written down, or forgotten, is not yet a fact anyone can report. It's a bet, being made in real time, by companies that are simultaneously one another's investors, customers, and rivals, precisely because none of them can yet say with confidence which side of that bet they're on.

Some of the money moving through this circle, and an increasingly consequential share of it, is now flowing one layer further down than the cloud contracts and model equity this chapter has traced, down into the physical business of manufacturing the chips themselves. That's where the fork leads next.

Chapter 9

The Substrate

> Globalization, at least in the chips sector, is dead.
>
> —Morris Chang, founder, TSMC

There was a time during the late 1990s and through much of the 2000s when, frankly, hardware and semiconductors were just not sexy. Except for a dedicated cadre of people who really geeked out on designs and power, the silicon that ran our computers was nearly faceless. Yes, Qualcomm made great strides in mobile, and ARM, more than anyone, finally broke Intel's lock on personal computers and servers. But that all changed with the advent of AI. Semis are back and in a big, strategic way. But the silicon industry is like a giant funnel with all the business dropping into the lap of two companies.

Trace any dollar from the previous chapter's financial circle back far enough, and it eventually arrives at the same place. Nvidia's chips, Broadcom's custom silicon, AMD's accelerators, Google's TPUs, even the Ironwood chips Google leases to Anthropic, all of it gets physically manufactured in one of a small number of factories clustered on a single island a hundred miles off the coast of mainland China. That stays true regardless of whose logo is on the box or whose balance sheet financed the deal. The company that owns those factories is Taiwan Semiconductor Manufacturing Company (TSMC). If this book has been searching for the cleanest possible instance of the Indispensability Trap across five build-outs, no need to look any further than this one.

TSMC controls somewhere between 65 and 72 percent of the global pure-play foundry market, depending on whose count you use. Samsung is a distant second at roughly 7 to 13 percent and nobody else is even in the same conversation. But even TSMC's dominant figure alone understates the true concentration, because it blends TSMC's older, commodity-grade manufacturing lines with the leading-edge processes that actually matter for AI.

Narrow the lens to advanced chips, the 7-nanometer and smaller processes that power every frontier accelerator we've discussed, and TSMC's share climbs past 90 percent. When Donald Trump told a crowd in 2025 that TSMC controlled 97 percent of the market, he was exaggerating the headline number, but not by nearly as much as his critics wanted him to be. On the metric that determines whether the AI build-out can proceed at all, the exaggeration was close to true.

TSMC generated $122.4 billion in revenue in 2025, up nearly 36 percent from the year before, at a net margin above 45 percent, a level of profitability that would be extraordinary for a software company and is almost unheard of in heavy manufacturing. Its client list is, effectively, the entire fabless semiconductor industry: Apple, Nvidia, AMD, Broadcom, Qualcomm, and dozens of smaller designers who have never operated a fab and have no plan to. Even Intel, a company that owns and operates its own manufacturing plants, sends some of its most advanced chip designs to TSMC because Intel's leading-edge process has, for most of the last decade, lagged TSMC's. Every company this book has profiled as a builder rather than an owner of infrastructure, going back to the cloud-era winners, is in this specific respect a tenant of a landlord it does not control and, at least for the moment, cannot replace.

The one company positioned to be a genuine second choice has, if anything, gone in the wrong direction. Samsung's foundry share peaked at roughly 15 percent earlier in the decade before sliding into the distant-second position noted earlier. It was undone by yield problems on its own 3-nanometer process that pushed customers back toward TSMC even when Samsung's prices were lower. A near monopoly usually invites a well-funded challenger eventually. Here, the best-funded challenger spent the years it might have used to close the gap watching the gap widen instead.

The fabless model itself explains why nobody has simply built around TSMC. Designing a chip and manufacturing one are almost entirely separate disciplines. The manufacturing side has, over five decades, concentrated into an industry where accumulated process knowledge, not equipment, is the barrier. What cannot be purchased is the tacit knowledge, the accumulated years of tuning yields, training technicians, and solving problems nobody has written down anywhere. This allows TSMC to turn advanced machinery into chips that actually work at scale and at a profitable cost.

Building a new leading-edge fab from scratch takes three to four years and more than $10 billion even before accounting for the learning curve. Intel's attempt to catch up, a $100 billion commitment to new plants in Ohio, has been repeatedly delayed, with the first facility not expected to open until 2030, the same year TSMC expects its Arizona expansion to be substantially complete. Nobody is closing this gap on anyone else's schedule but TSMC's.

The Company Behind the Company

TSMC's monopoly rests on a supplier whose monopoly is even more complete. The lithography machines capable of printing features a few atoms wide, the tool without which none of TSMC's leading-edge capacity exists, are built by exactly one company on Earth. ASML, based in a small Dutch town called Veldhoven. Where TSMC controls somewhere north of two-thirds of its market, ASML controls all of its market.

No competitor makes a machine capable of extreme ultraviolet lithography at any price. The machines took roughly three decades and several billion euros of research to develop. They involve components from thousands of specialized suppliers worldwide, including mirrors from a German optics firm polished smoother than almost anything else humans manufacture. They sell for well over a $100 million apiece, with the newest High-NA variants costing considerably more. TSMC, Samsung, and Intel all depend on ASML completely. Commentators have taken to calling ASML the new OPEC of the chip industry, though the comparison undersells the point. OPEC's members control roughly 40 percent of world oil production between them. ASML alone controls all of the market that matters most.

Even here a hedge is being built. A US-backed startup called xLight, chaired by former Intel chief executive Pat Gelsinger, was assembling roughly half a billion dollars by mid-2026, between a $150 million commitment from the CHIPS Act and a $350 million Series B round underway. They are seeking to develop an alternative light-source technology that could eventually loosen ASML's grip on the industry's most critical tool. The round is small relative to what a genuine competitor to ASML would require, and a working alternative remains years away by any realistic estimate. But the instinct behind it, that a single point of failure this consequential cannot be allowed to sit entirely outside American control, is the same instinct that produced the Intel rescue examined later in this chapter, which arrives one layer further up the supply chain and is considerably further from being solved.

The Chokepoint Inside the Chokepoint

But that's not the end of the monopolies. TSMC's monopoly has another monopoly nested inside it. Fabricating a working silicon die is only half of what it takes to produce a modern AI accelerator. The other half is advanced packaging, the process of physically bonding a logic die to the high-bandwidth memory stacked beside it. Thus, allowing the two to exchange data fast enough to be useful for chips at the frontier of AI performance. The only version of that process that works at a commercial scale is TSMC's CoWoS, short for chip-on-wafer-on-substrate. A perfectly fabricated chip that never gets packaged is not a product. It's a very expensive paperweight.

TSMC's CoWoS capacity sat at roughly 35,000 wafers a month at the end of 2023. It reached about 75,000 by the end of 2024 and was targeted to hit somewhere between 120,000 and 130,000 by the end of 2026, nearly a fourfold expansion in three years. Industry trackers describe it as still not enough. Nvidia alone has reserved the majority of that capacity through 2027.

TSMC's chief executive, C.C. Wei, told shareholders at the company's June 2026 annual meeting that CoWoS capacity remained extremely tight and was sold out through 2027. Nvidia's management has said the same in its earnings calls. Pricing tells the same story more quietly. Advanced packaging costs have been rising at two to four times the rate of the underlying wafers, which means the scarcity isn't really in the silicon anymore. It's in the glue holding the silicon together.

Intel Foundry offers a plausible alternative packaging technology of its own, called Foveros, and it may eventually offer real relief. As of 2026 it remains unproven at the scale and complexity the AI industry actually needs. A reminder that even where an alternative to TSMC is being actively built, as it is at almost every layer examined in this chapter, it exists first as a hedge and only later, if ever, as a genuine substitute.

The Silicon Shield

Taiwan produces roughly 60 percent of the world's semiconductors and, again, more than 90 percent of the most advanced ones, almost entirely through TSMC. This has given rise to a term used seriously by strategists on all sides of the Taiwan Strait: the silicon shield. The theory holds that Taiwan's indispensability to the global economy functions as a deterrent against invasion in its own right, independent of any formal military alliance. China, on this theory, cannot afford to disrupt an industry its own economy depends on. The United States and its allies, dependent on the same chips for everything from smartphones to fighter jets to the AI build-out, have a standing incentive to prevent that disruption by whatever means available. One widely cited estimate puts the global economic cost of a serious disruption to Taiwan's chip output at something like $2.5 trillion a year. Indispensability, in this specific case, is not just a mechanism that caps a company's returns. It is being treated, by serious people in national security establishments on multiple continents, as a literal deterrence architecture.

The shield is not static, and by 2026 there was open debate about whether it was eroding. Some of the erosion is TSMC's own doing. Facing simultaneous pressure from Washington to reshore production and from its own board to diversify geopolitical risk, TSMC has committed roughly $165 billion to new fabrication and packaging capacity in the US. The largest single direct foreign investment in American history, covering six wafer fabs, two advanced packaging facilities, and a research center on a campus outside Phoenix, Arizona, that TSMC expects will eventually support tens of thousands of permanent jobs on top of the roughly 40,000 construction jobs the build-out requires. Its first Arizona fab has been in volume production of 4-nanometer chips since late 2024, supplying Apple and Nvidia among others.

The second is targeting 3-nanometer production in the second half of 2027. A third, aimed at 2-nanometer capacity, isn't expected until 2028 or 2029, at which point Taiwan's fabs will likely already be running the next node behind that. The Arizona fab is real, and it is not small. It is also permanently a step behind the island it was built to hedge against. That diversification is precisely what a company in TSMC's position would do to protect its own business continuity. It is also, from Taiwan's strategic perspective, a slow unwinding of the one asset that makes an invasion of Taiwan riskier for China than it would otherwise be.

TSMC has been careful to keep its most advanced processes, the 2-nanometer node now and the 1.4-nanometer node coming after it, on the island itself, with only older, though still highly capable, processes running at the Arizona site. Whether that distinction is enough to preserve the shield's deterrent value or whether it's a distinction that matters less each year the overseas fabs mature is a live argument among people who study this for a living.

The military calculus sitting underneath that argument has its own timeline. Chinese military planners have reportedly targeted 2027 as a benchmark for being ready to act on Taiwan if ordered to. December 2025 saw exercises around the island that observers described as the largest and most sustained Taiwan-focused drills the People's Liberation Army had yet conducted. Manifold Markets, as of early 2026, implied something like a 22 percent probability of a serious cross-strait conflict by 2027, a number several analysts flagged as strangely low relative to how casually it's discussed in policy circles, and lower still than the risk premium one might expect given how much of the global economy runs through the same hundred-mile stretch of water.

Some of the more sobering recent analyses have moved past the invasion scenario entirely and toward a quarantine: China using its coast guard to inspect and delay shipping in and out of Taiwan's ports, choking the island's economy without ever needing to seize a single fab. TSMC itself has stated, more than once, that its facilities would become inoperable if an invasion were attempted, whether through deliberate shutdown or through the sheer impossibility of running an advanced fab without the constant, cross-border coordination of equipment servicing, materials supply, and software updates that a wartime blockade would sever instantly. This may be the only piece of critical infrastructure examined that was, in effect, built to destroy its own value rather than let itself be captured.

China's Answer

The US has been tightening export controls on advanced chip technology to China since 2018, and sharply since October 2022. That is when the Biden administration cut off access to the most capable chips, the design software needed to create competitive alternatives, and the manufacturing equipment needed to produce advanced chips domestically. The stated goal was straightforward: slow China's progress in AI and advanced computing by denying it the tools to build a competitive semiconductor industry of its own. The controls have, by most honest accounts, done exactly that, at least at the frontier. They have also done something the people who designed them did not fully anticipate. They convinced China's government that the only durable answer to a chokepoint it doesn't control is building one it does, at almost any cost.

The results are uneven, and that unevenness is itself the story. On mature-node equipment, wafer substrates, and legacy chip production, China's domestic industry has made fast, well-documented progress. Local suppliers are capturing meaningful share in a market Western and Japanese firms dominated only a few years earlier. On the frontier, the picture is starkly different. SMIC, China's largest foundry, is still dependent on aging deep ultraviolet lithography tools from ASML. The tools are pushed well past their intended purpose through multiple exposure passes to produce chips that are usable but nowhere close to TSMC's efficiency or cost structure. A domestic extreme ultraviolet machine, the kind of tool that would let China skip the workaround entirely, remains in prototype, with functional output realistically years away by most outside estimates. China's semiconductor plan, reported in early 2026, set an 80 percent self-sufficiency target and a 2030 goal for a fully domestic 7-nanometer production line. Interestingly, a target that echoes an almost identical, and substantially missed, goal China set for itself a decade earlier under Made in China 2025.

None of this changes TSMC's position at the frontier in the near term. What it does is confirm, from the other direction, is the same lesson Intel's rescue teaches from the American side: When a chokepoint this consequential can't be regulated the ordinary way, because it sits outside anyone's jurisdiction but its own, the response isn't patience. It's an expensive, often uneconomical race to build a second one, running in parallel on both sides of the Pacific. Each side betting that indispensability this total simply cannot be allowed to remain someone else's alone.

The chokepoint cuts in both directions at once, which is easy to miss if you only watch the export control side of the ledger. In early 2026 the same administration that had converted CHIPS Act grants into an Intel stake also approved the sale of Nvidia's H200 chip to Chinese customers, charging a 25 percent government cut on every unit exported. Beijing's answer was not to buy them. Chinese customs authorities blocked the shipments outright, and by the middle of 2026, by the US government's account, not a single H200 had actually been sold into China. This even as Chinese firms, including Alibaba and ByteDance, had reportedly lined up to order hundreds of thousands of units. Trump himself, briefing reporters after a summit with Xi Jinping, said China had simply "chosen not to" approve the purchases because "they want to develop their own." Nvidia eventually halted H200 production for the China market entirely and reallocated the TSMC capacity to its next-generation architecture instead.

In the meantime, Huawei's Ascend chips were the visible beneficiary: DeepSeek ran inference for its latest model on Huawei silicon rather than Nvidia's, and Nvidia's share of the Chinese AI chip market, by Jensen Huang's own account, dropped to zero. Access, once treated in Washington as a privilege to be sold, was refused on the other end as a dependency not worth the price of admission. We'll pick up this thread when we turn to China's AI build-out in the next chapter.

The Rescue

The state cannot regulate TSMC's returns the way Vanderbilt's railroads or Vail's telephone network eventually were because TSMC sits entirely outside its jurisdiction. This means that the access-guarantee mechanism has to find a different lever. In the American case, that lever turned out to be building a domestic second source, whether or not that second source could ever fully substitute for the first.

The company chosen for that role was Intel, and the path there began with a rumor. In the spring of 2025 reports circulated that a consortium of Intel's customers—Nvidia, AMD, and Broadcom among them—was exploring a joint takeover of Intel's struggling foundry business. TSMC itself was reportedly involved in some version of the discussions. Whatever the accuracy of the details, the prospect of Intel's manufacturing arm, the closest thing the US has to a domestic leading-edge fab, ending up under the operational influence of the very foreign company the government was trying to reduce dependence on appears to have concentrated minds in Washington considerably.

What emerged instead, in August 2025, was a direct nationalization of a slice of Intel itself. The US Treasury converted existing CHIPS Act grants into a roughly 10 percent, nonvoting equity stake in the company, worth on the order of $8.9 billion. That made the federal government Intel's single largest shareholder. Commentators at the time called it, half-mockingly and half-seriously, a Silicon Sovereignty deal, and it did not stand alone. Around the same period, the administration extracted a comparable arrangement from Nippon Steel in exchange for approving its acquisition of US Steel, a nonequity golden share giving Washington a say over specific corporate decisions. Trump himself floated the idea of taking equity positions in other strategically sensitive companies, defense contractor Lockheed Martin among them. Intel's stake was not a one-off improvisation. It was the clearest instance, so far, of a pattern this administration was prepared to repeat wherever it judged a private company's fortunes to be entangled with the state's own strategic interests.

Intel's chief executive, Lip-Bu Tan, had spent much of 2025 in direct, personal negotiation with the administration, at one point publicly addressing the president's stated doubts about his leadership after reports surfaced questioning his past business ties to Chinese firms. That Tan emerged from the year with the government as his largest shareholder, rather than as a casualty of the doubts surrounding him, shows just how entangled Intel's survival had become with the state's industrial strategy by that point. Congressional reaction split along predictable lines, with critics on both the left and the libertarian right. They warned that a government equity stake in a publicly traded company, acquired without the ordinary legislative process a nationalization would typically require, set an uncomfortable precedent regardless of which administration held it. The precedent held anyway.

Weeks later Nvidia added its own weight to the arrangement, investing $5 billion for a stake reported at roughly 4 percent, alongside a partnership to codevelop custom x86 chips integrating Nvidia's NVLink interconnect technology. This wasn't purely a political favor. Nvidia's Blackwell generation of AI accelerators had run into real thermal and advanced packaging problems earlier in 2025. Intel's packaging capacity, largely idle amid the company's broader struggles, offered a plausible fix that happened to align with what the administration wanted anyway. The Federal Trade Commission cleared the investment in December 2025, and by early 2026, Intel's 18A manufacturing process was reporting yield improvements of roughly 7 percent a month. A pace that, if it holds, would put a real domestic alternative to TSMC's most advanced nodes within reach for the first time in years. An even more advanced 14A node is targeted for risk production later in 2026.

None of this makes Intel a genuine substitute for TSMC yet, and nobody involved claims it does. What it does is establish, in real time and one layer beneath the cloud and model layers already examined, the exact access-guarantee pattern traced in every prior build-out. It is arriving earlier and more explicitly than in any of them. The railroads got rate regulation only after decades of documented abuse. The Bell System's universal service bargain took the credible threat of a hostile Justice Department. Here, the government didn't wait for TSMC to abuse its position, largely because TSMC, sitting outside US jurisdiction, was never going to be regulated the conventional way in the first place. It moved instead to manufacture leverage of its own, by buying a stake in the one domestic company positioned to someday offer an alternative.

The same administration reached for an adjacent lever within weeks of the Intel stake. Striking an unusual arrangement in which Nvidia and AMD agreed to remit 15 percent of their China chip sale revenue to the US government in exchange for export licenses to sell restricted chips into that market at all. Critics, including analysts at the Cato Institute, warned that a government sitting simultaneously as Intel's largest shareholder and as the gatekeeper of export licenses and tariff relief for Intel's competitors had built itself real leverage to eventually pressure those competitors into routing manufacturing toward Intel, whether or not Intel's technology justified the choice on its own merits.

That is the Indispensability Trap's strategic capture mode again. The same mode this book will examine in full when it reaches the export control episode that briefly suspended Anthropic's Fable and Mythos models in June 2026. Here it shows up one layer further down the stack. Aimed not at a model but at the physical silicon everything else depends on, and it shows up before the crisis rather than after it. That is its own kind of evidence for how thoroughly the pattern has been internalized by the people running this build-out the second time around.

Fabless and Exposed

Nvidia and AMD occupy a position in this chapter that neither company would have chosen for itself a decade ago: Both are essentially design houses with no meaningful manufacturing capacity of their own. Both are entirely dependent on TSMC for the leading-edge silicon their businesses are built on. Both have now taken equity positions in Intel, a company they compete against in adjacent markets, partly because the alternative to hedging their single point of failure was accepting it indefinitely.

Nvidia's exposure runs in two directions at once. It needs TSMC for the accelerators that generate the overwhelming majority of its revenue, and it needs Intel, or at least a credible prospect of Intel, as insurance against exactly the geopolitical scenario the previous section described. AMD's position is close to identical, and AMD was reportedly part of the same consortium of customers whose interest in Intel's foundry assets helped trigger the government's intervention in the first place.

This is the instability principle recurring at a layer even more fundamental than the cloud and model layers already examined. Google could not hold still as the grid's owner, tenant, and landlord simultaneously. Microsoft could not hold still as OpenAI's investor, Anthropic's customer, and a competitor to both. Now Nvidia and AMD cannot hold still either, hedging against the single most consequential chokepoint in their entire industry by taking a stake in a company whose own manufacturing technology still lags the company they're actually dependent on. The hedge is not a vote of confidence in Intel. It is an insurance premium against Taiwan, priced and paid years before anyone can say whether the policy will ever need to pay out.

Broadcom belongs in this picture too, and its presence here closes a loop opened in the last chapter. Every guarantee Broadcom wrote for Anthropic's TPU leases and every dollar it advanced ahead of OpenAI's contribution to Project Nexus sit on top of custom chips that Broadcom itself does not manufacture. Broadcom is fabless in exactly the same sense that Nvidia and AMD are, dependent on TSMC for the physical silicon underneath every guarantee it issues. The company willing to put its own balance sheet behind its customers' chip purchases is itself standing on a foundation it does not own and cannot control. The guarantee was never fully Broadcom's to give. Some part of it belonged to TSMC all along.

What the Substrate Reveals

Every engine explored throughout the previous build-outs applies to TSMC with unusual clarity, precisely because TSMC sits at a point in the stack where the ordinary version of each engine breaks in an instructive way. The Indispensability Trap predicts that extreme essentiality invites a capped return, through rate regulation, antitrust, or strategic capture. TSMC is about as indispensable as any company examined in this book, and yet its own returns remain, for now, entirely uncapped, precisely because the country best positioned to cap them has no jurisdiction to do so. The trap hasn't failed here. It has redirected. Unable to reach TSMC's balance sheet, it has reached instead for Intel's, for Nvidia and AMD's China revenue, and for the export control apparatus that will resurface later as the most direct lever of all.

Jevons paradox is visible in the sheer scale of demand TSMC's efficiency has unlocked. More chips at falling per-unit cost driving more AI capability, which drives more demand for chips in a loop with no natural ceiling in sight. It's visible again in the CoWoS build-out, where a fourfold capacity increase in two years still isn't enough. The Perez framework's installation frenzy, the same frenzy traced through AI's circular financing, is visible here too. Hundreds of billions of dollars of fab and packaging construction racing to catch up with a demand curve nobody can yet confirm is real, one node behind the frontier and several years behind the money already committed to it.

What TSMC's position ultimately demonstrates is the fork posed at the start of Part III. Whether owning the full stack finally lets an infrastructure owner keep what it built doesn't resolve cleanly even at the very bottom of that stack. TSMC owns the single most indispensable layer in the entire AI build-out, more indispensable than the cloud, more indispensable than any model, and its response to that position has been to diversify away from the very geography that makes it indispensable in the first place. Meanwhile the country most dependent on it builds a hedge against needing it at all, and the country most locked out of it races to build a third option nobody outside China particularly wants to see succeed. Indispensability, at this depth, doesn't look like power. It looks like exposure, managed as carefully and as anxiously as every other kind of exposure traced here, one layer further down than anyone expected to need to look.

The commoditization we've followed from the grid to the cloud to the physical chips themselves has one layer left to climb before it reaches the place everyone originally assumed was safe. That's the model itself, and that's where the fork leads next.

Chapter 10

The Free Model

> Information wants to be free. Information also wants to be expensive.
>
> —Stewart Brand, Hackers' Conference, 1984

Every build-out we have examined eventually reached a layer that resists commoditization, at least for a while. Railroads gave way to freight brokers who arbitraged rates between them. Telephone equipment gave way to the long-distance network effects Vail built on top of it. Electrification gave way to the appliance, engineered to run on standardized current that any utility could supply. The cloud era gave way to the platforms that treated bandwidth and storage as a free input and built where the commodity was cheapest.

In each case, something sat one layer above the commodity and held its value. AI's twist is that this time the commodity wave doesn't stop. It has now reached the layer everyone, including the labs building it, assumed would be the one place value could not be dislodged: the frontier model itself.

This chapter is about another arrow in the Indispensability Trap's quiver: open source. I watched this movie the first time it ran. Open source went from a curiosity in the late 1990s to, over the decade after 2006, the default way software gets built and licensed, and it turned the industry's entire revenue model upside down along the way. The companies that survived the wave did it by selling something one layer above the free code, the hosting, the support, the running service. This is how open source ended up powering the SaaS economy instead of killing the software business.

The AI model market is proving no different. The question is whether open models are genuinely challenging the closed frontier, and the answer arrived while I was finishing this book.

Stanford's 2026 AI Index measured the gap in March. The best closed model led the best open model by 3.3 percent on the index's composite measure. This was in fact wider than the 0.5-point gap of August 2024, a reminder that the race has tightened and loosened and tightened again rather than closing in a straight line. Six of the top ten models on the widely watched Arena leaderboard were closed, and every open model in the top tier was chasing from below.

Then came eleven days in July. On the sixteenth Moonshot AI released Kimi K3, a 2.8-trillion-parameter model with a million-token context window. It's the largest open-weight model ever announced; its full weights committed for public release by month's end. On the independent Artificial Analysis rankings, it landed above one of the closed frontier flagships, and in blind testing developers preferred it over every leading American model for front-end coding. Three days later, Alibaba released Qwen 3.8 tMax, 2.4 trillion parameters aimed squarely at coding and agentic work. Alibaba claimed it trailed only a single closed model on capability. By late July, depending on which leaderboard you consulted, close to half of the top dozen models in the world were Chinese, and most of those were open weight.

A model does not need to win every benchmark to change the market. It needs to be close enough, cheap enough, and open enough that owning it beats renting something marginally better.

The number that matters more than the gap itself is how fast the gap has been closing. Epoch AI, which tracks this as a time lag rather than a percentage, found that open-weight models trailed the closed-source frontier by roughly three months on average in early 2026. In late 2024, that same lag was closer to a year. After K3, analysts who had been debating whether the lag was six months or nine were revising it down to a matter of weeks.

On raw knowledge benchmarks like MMLU, the picture had gone further still. The gap that stood at nearly eighteen points at the end of 2023, the best closed model scoring around 88 percent against the best open model's 70.5, had effectively closed to zero by early 2026. On coding, the practical benchmark that determines whether a company can build its product on an open model instead of paying for a closed one, the story is now a procession.

DeepSeek's V4-Pro, released in April 2026 under an MIT license, scored

80.6 percent on SWE-bench Verified, the highest open-weight result on the benchmark at that time, while charging a fraction of frontier prices. Zhipu AI's GLM-5.2, released in June with MIT-licensed weights, beat GPT-5.5 outright on long-horizon coding benchmarks at roughly one-sixth the cost. Then the July wave, K3 and tMax back to back put open models at the top of the coding leaderboards themselves. They sport token prices that turn a fifty-million-token monthly workload from over a thousand dollars into a couple hundred.

Within days, American AI companies were calling on Washington to regulate Chinese open models, which told you the stakes better than any benchmark could. And it clarified what game is actually being played. The US is competing to build the world's best AI. China is competing to become the AI the rest of the world builds upon. Those are not the same contest. China is not stumbling into the Indispensability Trap at the model layer. It is springing the trap deliberately, commoditizing intelligence before American companies can convert a temporary lead into permanent control.

The most telling evidence that even the most committed closed model labs no longer believe the model layer is defensible on its own came from OpenAI. In 2025 the company most associated with keeping its weights locked released gpt-oss-120b, a genuinely open model under a full Apache 2.0 license, competitive enough on competition coding benchmarks to match or exceed OpenAI's smaller closed models. A company built almost entirely on the premise that its frontier models were worth paying for chose, on its own initiative, to give one away for free. That is not the action of a lab that believes the model itself is where its future value will be defended.

Not every lab facing the identical pressure has reached for the identical response, and the divergence is itself informative. Google released its Gemma line of smaller models under Apache 2.0 while keeping its flagship Gemini models entirely closed, hedging at the low end without conceding the frontier. Meta went open with Llama but attached a community license restrictive enough that it isn't recognized as open source by the standard industry definitions. But it does capture some of openness's adoption benefits while retaining more legal control than Mistral or DeepSeek chose to keep. Anthropic, alone among the major American labs, has released no open-weight model at any scale, betting that its defensible ground lies entirely above the model layer rather than in any hybrid position within it. Four labs facing the same commoditization pressure produced four different strategic answers, ranging from full capitulation to full resistance, which suggests nobody in the industry, including the people building these models for a living, has worked out which position is the right one to hold.

How This Happened

Three forces converged to produce this, and none of them required a single company to make a mistake for it to happen.

The first is architectural convergence. By 2026 nearly every flagship model at real scale—Mistral's Large 3, Qwen3-Coder, DeepSeek's V3 and V4 line, Kimi K2, Zhipu's GLM series, Meta's Llama 4 Maverick—had independently arrived at the same underlying design. A sparse mixture-of-experts network, where only a fraction of the model's total parameters activate for any given token, delivering the quality of a much larger dense model at a fraction of the inference cost.

Nobody copied this from a single source. Labs in California, Paris, and Beijing converged on it separately, the way competing nineteenth-century railroads all eventually converged on standard gauge. Running anything else got too expensive to sustain. When independent teams working with different data, different funding, and different incentives keep arriving at the identical architecture, that's a signal the architecture has become the obvious answer rather than anyone's proprietary edge.

The second force is that the biggest remaining lever, more data and more compute thrown at ever-larger dense models, has been running into diminishing returns for over a year. This pushed every lab toward efficiency gains instead. DeepSeek supplied the clearest public demonstration of what an efficiency-first approach can do. Its V3 base model was trained using roughly 2.6 million GPU hours, against the 30.8 million GPU hours Meta's Llama 3 405B had required, an order-of-magnitude difference in training efficiency between two models in the same general capability class. Techniques like this, once published in a paper or reverse engineered from a released model's behavior, are not proprietary for long. They become the new floor every other lab is expected to clear.

The third force is licensing itself, and here the story gets more interesting than a simple race to give everything away. Apache 2.0, the most permissive major license in wide use, has become the default for most Western open-weight releases, Mistral's entire model family and Google's Gemma line among them. But the Chinese labs racing hardest on capability have not converged on maximal openness the way the architecture question converged on mixture-of-experts. Some, like DeepSeek, ship under MIT, comparably permissive. Others have pulled in the opposite direction: MiniMax shifted its M2.7 model from a fully open MIT license to a noncommercial license requiring written authorization for commercial use. A stark reminder that once a lab has captured the adoption and attention that openness buys, nothing obligates it to keep giving the model away on the same terms indefinitely. Commoditization at this layer is not a one-way ratchet. It's a live negotiation, refought release by release, between how much openness a lab needs to win adoption and how much it can claw back once adoption is secured.

The Bet That Paid Off

Nowhere is the case for embracing rather than resisting this commoditization made more concretely than at Mistral. The French lab's flagship release, Mistral Large 3, arrived in December 2025 as a 675-billion-parameter mixture-of-experts model, trained from scratch on three thousand Nvidia H200 GPUs. It was released under full Apache 2.0 alongside a family of smaller Ministral models carrying the same license. This wasn't a company begrudgingly giving up ground. It was a full, deliberate bet that giving away the model itself was the better business than trying to protect it.

The bet appears to be working. Mistral's annualized revenue grew from roughly $20 million to $400 million in the space of a single year, a twentyfold increase. The company's valuation reached $13.8 billion, with an $830 million raise in March 2026 earmarked for new data center capacity near Paris and in Sweden. None of that revenue comes from charging for the weights themselves; those are free to anyone who downloads them. It comes from everything built one layer above the free model. Products and services such as Le Chat, Mistral's consumer and enterprise chat product; Forge, an enterprise platform for training custom models on proprietary data; Voxtral, a voice model undercutting ElevenLabs on price; and the cloud infrastructure and support contracts that come with running any of it at scale.

Even Mistral's smaller open releases carry real competitive bite, the fourteen-billion-parameter Ministral reasoning variant scored 85 percent on the AIME mathematics competition, beating Alibaba's comparably sized Qwen-14B outright. Further evidence that giving a model away doesn't require making it second-rate.

Mistral gave away the layer every other kind of infrastructure owner has been fighting to protect. In the meantime, it built its actual business in the layer just above, which is precisely the pattern this book's engines predict once a resource becomes cheap enough. This is Jevons paradox in its purest recent form, freer models driving more total usage, more total usage driving more revenue for whoever captures the layer built on top of it, even when that layer belongs to the same company that gave the model away.

One research group tracking the closing gap between open and closed frontiers, extrapolating the trendline forward, projected that the two would converge entirely by early December 2026, a claim its own authors flagged as more a measure of momentum than a firm prediction. The trendline assumes today's rate of improvement holds steady on both sides. Whether or not that specific date proves right, the direction of travel it's extrapolating from is not in serious dispute.

Cheaper and More of It

Model capability converging is only half the commoditization story. The other half is what's happened to the price of using any of these models at all. It is the clearest real-time demonstration of Jevons paradox we've encountered since the fiber glut. Epoch AI's research tracked the price of reaching a fixed performance milestone, not the price of a model in the abstract, and found declines ranging from roughly ninefold to as much as nine-hundred-fold per year depending on the benchmark, with the steepest drops concentrated in the period after January 2024.

The plainest illustration: The level of performance GPT-3 offered on general knowledge questions in November 2021 at a cost of $60 per million tokens was available by March 2026 for six cents per million tokens or less, a thousandfold collapse in under five years. Andreessen Horowitz coined a name for this deflationary curve, LLMflation, explicitly comparing it to Moore's Law. Enterprise spending trackers captured the same trend on a shorter horizon, the average cost per million tokens across major providers fell from roughly $10 to $2.50 in a single year.

Jevons's nineteenth-century observation was that cheaper coal didn't reduce Britain's total coal consumption, it multiplied it by making uses of coal economical that hadn't been worth the expense before. Cheaper tokens have done exactly this to AI usage. Reasoning models, which generate thousands of internal tokens working through a problem before producing a final answer a fraction of that length, can burn ten thousand tokens of invisible computation to answer a question a 2022-era model would have answered directly.

Always-on agentic workloads, software that monitors email, logs, and market data continuously rather than responding to a single prompt, consumed essentially no compute in 2024 but had become a meaningful share of enterprise inference budgets by 2026. The result is a paradox stated plainly by more than one enterprise buyer: Per-token prices fell by three-quarters year over year, but the total bill still went up because the average enterprise AI budget grew from roughly $1.2 million in 2024 to $7 million in 2026, even as the unit price of intelligence kept collapsing underneath it.

OpenAI's financials are the sharpest illustration available of what this paradox does to a lab's balance sheet even as it drives adoption. The company generated roughly $13 billion in run-rate revenue by late 2025 against losses estimated in the range of $8 to $12 billion, spending well over a dollar for every dollar of revenue it brought in, a dynamic inseparable from the circular financing arrangements I've already examined in detail.

Alphabet's capital expenditure guidance tells the same story from the supply side. Projected at $75 billion for 2025, it was revised to a range of $180 to $190 billion for 2026, raised at Q1 2026 earnings from an earlier $175 to $185 billion range at the start of the year, nearly tripling in a single year. This was overwhelmingly directed at the infrastructure needed to serve a demand curve that cheaper tokens themselves helped create. Falling prices and rising infrastructure spending are not in tension. They are the same mechanism viewed from opposite ends of the same transaction.

The free models at the center of this chapter are not bystanders to this price collapse. They are one of its principal causes. Once a capable model is available under a license that costs nothing, every closed competitor is forced to price its own offering against a free floor, whatever margin it might otherwise have preferred to hold onto. Model commoditization and the collapse in the price of intelligence are not two separate trends running in parallel. They are the same force, described from two different vantage points.

The Uptake at Scale

None of this matters if nobody actually uses the free models in production, and by 2026 the usage data left no ambiguity on that question. By the second quarter of 2026, Chinese open-weight providers alone accounted for more than 45 percent of all traffic on OpenRouter, the widely used routing layer that lets developers switch between models. A single model, Xiaomi's MiMo V2 Pro, was moving nearly 4.8 trillion tokens a week, more than three times the volume of the next-largest model on the entire leaderboard. On coding specifically, MiMo V2 Pro and Alibaba's Qwen3.6 Plus together were routing something close to half of all coding tokens passing through OpenRouter, with MiMo V2 Pro's share running at roughly six times Anthropic's. Qwen's download numbers tell the same story from a different angle. Alibaba's model family passed 700 million cumulative downloads by January 2026, overtaking Meta's Llama in total downloads, on its way to becoming the most downloaded open-weight system in the world.

The clearest single data point on how deep this adoption runs came from Andreessen Horowitz. Martin Casado, a general partner at the firm, said publicly that among the roughly one in four startups pitching the firm with an open-source AI stack at all, there was an 80 percent chance they were running on a Chinese model rather than anything built in the US. A figure the US-China Economic and Security Review Commission corroborated independently in a March 2026 report drawing on Hugging Face's download data. One of the more pointed examples sits inside Cursor, the AI coding startup SpaceX acquired for $60 billion in stock in the deal examined in Chapter 8, which runs its Composer feature on Kimi K2.5, a Chinese open-weight model, rather than on a model built by any of the seven companies previously mapped as the players fighting over durable value in this build-out. One of the largest all-stock startup acquisitions in tech history, absorbed into the same company hedging its own GPU exposure through SpaceX, is itself built in part on a free model from Beijing. This is not a footnote. It's the commoditization argument closing its own loop inside a deal this book has already spent a chapter examining from a completely different angle.

Some of the response to this has taken the same shape as the hedges documented at the silicon layer. Reflection AI, a startup founded specifically to build a credible, well-resourced Western alternative to the Chinese open-weight wave, had reached an $8 billion valuation by 2026, funded on the same logic that produced xLight's bet against ASML's lithography monopoly. A single point of failure this consequential cannot be allowed to sit entirely outside American control, whether or not a working alternative exists yet. As with xLight, the money committed so far is real, and a credible substitute remains a work in progress.

The arithmetic behind the 80 percent figure is not complicated, and it's worth doing once in plain numbers rather than taking the trend on faith. A startup running a hundred million tokens a month through a frontier closed model at rates common in 2024 faced a bill in the neighborhood of $300,000. The same workload run through an open-weight model priced at a small fraction of that rate could cost under $2,000. For a company with a finite runway and an investor asking hard questions about burn rate, that isn't a preference between two roughly comparable options. It's the difference between a line item and an existential threat. Martin Casado's 80 percent is what happens when several thousand founders run that same spreadsheet independently and arrive at the same answer.

This directly complicates the map of the MAGOS 8 companies making eight different bets regarding where durable value would land. Google, Microsoft, Amazon, Meta, and Oracle were sorted as infrastructure owners; OpenAI and Anthropic as the pure-play model builders renting the grid; xAI as a hybrid of both. That map assumed the model itself was the asset worth fighting over. If a meaningful share of the world's actual AI usage is running on models that cost nothing to license, built by companies outside that map entirely, then the fight over which of the eight owns the winning model may be a fight over a shrinking share of the total pie. Genuinely durable value will be pushed even further up the stack than Chapter 7's original map anticipated.

Meta occupies an increasingly awkward position inside this story. It was Meta's original Llama releases that did more than any single company to legitimize open-weight models as a serious alternative to closed frontier labs in the first place. By 2026 Llama 4 Maverick remained a capable model, and its ecosystem of fine-tuning tools and community support remained the largest in the industry, mostly on the strength of two years of accumulated tooling inertia rather than current benchmark leadership. But Llama 4 ships under Meta's community license, which is not recognized as open source by the standard industry definitions and imposes usage restrictions on companies above a certain scale. That is a meaningfully less permissive stance than Mistral's unconditional Apache 2.0 or DeepSeek's MIT. The company that helped start this commoditization wave is now being outpaced on both licensing generosity and raw benchmark performance by labs that took the logic of Meta's strategy further than Meta was willing to go.

What Hasn't Commoditized, Yet

None of this should be read as a claim that every frontier capability has become interchangeable. Reasoning-heavy work remains the one place where closed models retain a real, if narrowing, edge. On the hardest benchmarks built specifically to resist saturation—GPQA Diamond, Humanity's Last Exam, frontier mathematics competitions—Claude, OpenAI, and Google's top reasoning models continued to outperform the best open alternatives by something like three to eight percentage points as of mid-2026. Google's Gemma line illustrated the shape of that gap concretely. Its 31-billion-parameter dense model scored a competitive 84.3 percent on GPQA Diamond against closed frontier leaders scoring around 94 percent, a real difference but not the chasm that separated open and closed models even two years earlier.

It is concentrated in exactly the workloads—the hardest multistep reasoning and the most demanding agentic tool use—that a lab's frontier researchers would identify as the genuine remaining edge of the art. For the vast majority of practical deployment—coding, summarization, customer support, data extraction, document analysis—the difference between an open model scoring 85 on a composite index and a closed model scoring 94 is, in Anthropic's honest internal assessment of the market, frequently invisible to the end user footing the bill.

That distinction, between frontier reasoning capability and practical deployed capability, is this chapter's real finding. The commodity wave has not reached the very tip of the frontier. It has reached everything just beneath it, which by volume and dollar value is most of what the AI industry actually sells. Whether the country most responsible for pushing the open-weight wave this far did so as an accident of engineering talent or as a deliberate strategy to win the world's developers away from American models entirely is a question that deserves its own full chapter, not a paragraph inside this one.

No Safe Rung

Put all of this together and the argument introduced at the beginning of Part III as a possibility becomes, by the middle of 2026, closer to an observed fact. Every previous build-out left a layer standing above the commodity precisely because building that layer required something the commodity itself couldn't supply: capital, network effects, physical infrastructure, brand trust. The frontier AI model was supposed to be that layer this time, defensible because training it required capital and talent at a scale only a handful of companies could muster. What the last eighteen months actually show is that once the architecture converges, once the efficiency techniques get published, and once even one well-resourced lab decides giving the model away is a better business than protecting it, the model itself becomes exactly as commodifiable as the compute and the chips beneath it.

All four of this book's engines converge on the model layer at once here, which hasn't happened as cleanly at any earlier layer in Part III. The Indispensability Trap predicts that something essential enough gets its returns capped, and the model layer has essentially capped its own returns from within, without a regulator or a court needing to intervene at all. A dozen labs racing to match each other's benchmarks accomplishes what an antitrust suit takes a decade to attempt. Jevons paradox is not a background metaphor here, it's the mechanism: cheaper tokens producing more total spending on tokens, exactly as cheaper coal produced more total coal burned. The Perez framework's installation frenzy explains why the capital keeps flowing even as the thing it's buying gets cheaper by the month. Labs and their backers are still betting on capturing a share of a market whose eventual size nobody can yet confirm, the same wager AI's circular financing traced one layer up the stack. The distinction between financial and production capital shows up here as the sharpest open question in the whole chapter. Mistral's twentyfold revenue growth suggests the free model strategy is building something durable, a genuine business layered on top of a commodity; OpenAI's $1.35-per-$1-earned economics suggest, at least for now, that some of this capital is still chasing a return the underlying business hasn't yet earned.

This doesn't mean nothing is defensible in the AI stack. It means the search for what's defensible has to look somewhere the commodity genuinely cannot follow. It's not model quality, which is converging faster than most forecasts anticipated, but proprietary data no competitor can access, distribution relationships built over years, workflows too embedded in a customer's operations to casually replace, and a trusted or regulated position that a cheaper alternative cannot simply copy by matching a benchmark score. That search is what the rest of Part III now turns to, starting with the country that has pursued open-weight commoditization more deliberately and for more explicitly strategic reasons than anyone else building at this scale.

Chapter 11

China's Two Loops

Necessity is the mother of invention.

—English proverb

On January 27, 2025, a Chinese AI lab most of the world had never heard of released a reasoning model called R1. Within days it had displaced ChatGPT as the most downloaded free application in the US App Store. American technology stocks fell sharply on the news. Nvidia alone shed roughly $600 billion in market value in a single trading session on the theory that a model built for a fraction of what frontier American labs were spending had just demonstrated the whole premise of the AI build-out might be wrong. The panic passed within weeks. What didn't pass was the pattern R1 had revealed: A Chinese AI ecosystem willing to give its models away for free, iterating faster in public than American labs were iterating behind closed doors, and doing it as a matter of deliberate, state-aligned strategy rather than as a scrappy startup's improvisation.

Marc Andreessen, watching Nvidia's stock chart from the venture capital side of Silicon Valley, called it AI's Sputnik moment within hours, reaching for the 1957 Soviet satellite launch that had shocked Americans into believing they'd fallen behind in a race they assumed they were winning. The comparison spread through the press within days, repeated by the *Guardian*, the *New Yorker*, and NPR, among many others, and it stuck for a reason. R1 hit the same nerve Sputnik had, the sudden, public revelation that a rival everyone had underestimated was closer than anyone had priced in. It also drew the same pushback a real Sputnik moment eventually earns. Researchers at CSIS and RUSI pointed out in the weeks and months that followed that DeepSeek hadn't actually leapfrogged American capability, that its cost claims omitted substantial development expenses, and that the technique it had refined, efficient training on constrained hardware, had been pioneered in American labs years earlier. Whether R1 was a true Sputnik moment or an overreaction dressed up in a convenient historical analogy is still contested, and reasonable people on both sides have a case. What both sides agreed on was that something had changed, and that the change was worth taking seriously either way.

I wrote about DeepSeek at the time and reached for an older analogy than Sputnik's satellite. Not the launch itself, but the design philosophy behind the Soviet space program that made the launch possible. Soviet engineers, working with a fraction of the material resources and manufacturing precision available to their American counterparts, had become good at building lighter, simpler, more ruthlessly optimized hardware precisely because they couldn't afford the alternative. Necessity, in that program, wasn't a slogan. It was the design constraint that produced the innovation. DeepSeek's team, denied the newest Nvidia silicon by the export controls we've already examined, did something structurally similar. They pruned, compressed, and algorithmically optimized their way to a result the best-funded labs in America had assumed only raw compute could buy. Resource scarcity didn't block DeepSeek's progress. It shaped the specific, efficient path that progress took.

The skepticism that followed the announcement was to be expected. In retrospect it was all over the place; not all of it aged the same way. Some of it was the ordinary, healthy kind. Analysts pointing out, correctly, that the widely repeated training-cost figure captured only the final run and left out years of prior experimentation, the AI equivalent of judging a drug's cost by the one clinical trial that worked while ignoring all the ones that didn't.

Some of it was darker and less resolved. Reporting at the time raised concerns that forced labor may have factored into the real cost of DeepSeek's infrastructure, a charge serious enough to warrant scrutiny and, as of this writing, still unproven. Some of the loudest skepticism came from people and institutions with an obvious stake in the outcome, such as established American labs and the hardware makers whose valuations depended on the assumption that frontier AI required exactly the kind of massive compute budgets DeepSeek had just shown could be worked around. None of that makes the skeptics wrong. It does mean the debate over DeepSeek's legitimacy was never purely a technical argument conducted by disinterested referees. Readers of any single hot take, including mine, were well served by remembering whose interests that take happened to serve.

Sitting underneath the panic was an irony that is frankly just too much to ignore. The open-source software movement, the idea that code should be shared freely, studied openly, and improved by anyone rather than locked behind a corporate license, is as American an export as any traced. It was born out of Unix hackers and free-software advocates, and it matured into the Linux, Kubernetes, and Apache projects that now run most of the internet's infrastructure. For decades that ethos was something the US held up as a natural extension of its own values—open information, open collaboration, open competition, all as expressions of a free society, contrasted against a one-party state that censored its own internet and jailed people for the ideas it didn't like.

By 2026 the positions had effectively swapped. The Chinese Communist Party, of all the world's governments, had become the loudest institutional voice arguing that AI should be an open, shared, freely distributed global public good. Meanwhile several of America's frontier labs, the direct descendants of the culture that built the open-source movement in the first place, were tightening access to their most capable models and charging by the token for the privilege of using them. Beijing did not stumble into this reversal. It calculated, correctly, that openness was the fastest available route to global developer mindshare, and it picked up a rhetorical position that the country that had invented it had seemingly set down.

By March 2026 the US-China Economic and Security Review Commission, the congressionally chartered body that tracks exactly this kind of strategic competition, had given the underlying pattern a name: two loops, reinforcing each other, and only one of them within reach of anything Washington had built to contain it.

The Two Loops

The first loop is digital, and it's what the previous chapter documented in its effects without naming its cause. Chinese labs release models under permissive licenses. Developers worldwide download and fine-tune them. The resulting improvements and use cases flow back into the ecosystem. The next generation of models starts from a stronger base than the last, all without any single lab bearing the full cost of the underlying research. The Commission's report described it as a diffusion-innovation feedback loop. Each model becomes raw material for the next, refined by a global community that neither Beijing nor any single Chinese company has to pay to employ.

The second loop is physical, and it's the one American policy has mostly missed. China's manufacturing base, the largest in the world, generates real-world interaction data at a scale no other country can currently match. Things like forklifts, assembly line robots, warehouse logistics, quality inspection cameras, all producing the kind of embodied, physical world training data that text scraped from the internet simply doesn't contain. Beijing has treated this as a formal strategic asset rather than an incidental byproduct, designating data itself as a factor of production and permitting Chinese enterprises to carry data assets on their balance sheets the way they'd carry equipment or inventory. One cited example, a factory in Guangdong outfitted with 5G-connected cameras, uses the resulting data to cut equipment repair rates by 20 percent and save more than a million renminbi a year. A modest number in isolation but multiplied across a manufacturing base the size of China's, it becomes the shape of a national training data set that no Western economy currently has a way to build.

The Commission's framing of why this matters is worth quoting close to verbatim, because it states the strategic asymmetry more precisely than any paraphrase could. American export controls target the digital loop, restricting the advanced chips needed for frontier model training, but they are not well suited to reaching the physical loop of deployment-driven data accumulation across China's factories and supply chains. Washington built a chokepoint for silicon. It did not, and arguably cannot, build one for the exhaust data of an entire manufacturing economy.

None of this is just incidental to Chinese policy. It is the policy. Premier Li Qiang told the World Economic Forum in the summer of 2025 that China's innovation is open and open source. That the country is willing to share its indigenous technologies and innovation with the world, a line he repeated weeks later at the World Artificial Intelligence Conference in Shanghai. At the same event he also unveiled a thirteen-point Global AI Governance Action Plan and proposed a new World AI Cooperation Organization to help write the rules the rest of the world's AI development would follow. Beijing's AI+ Initiative, approved in full in August 2025 after more than a year of drafting, formally directs government support toward open-source AI ecosystems and sets targets for how many open models the industry should field at each tier of capability. This is not a description of what several Chinese companies happen to be doing. It is a national plan, with funding mechanisms attached, instructing them to do it.

The Digital Loop, in Practice

DeepSeek's history illustrates the digital loop from the inside. The company was founded and funded by High-Flyer, a Chinese quantitative hedge fund, with no public shareholders and no outside venture investors demanding a return on a fixed timeline. This structure lets it prioritize raw capability over near-term monetization in a way few Western labs can match. Liang Wenfeng, the fund's cofounder, made his money running trading algorithms years before he turned the same computing infrastructure toward frontier AI research. He has said little publicly beyond the research itself, a contrast with the Silicon Valley habit of narrating every model release as a leadership story.

R1's mixture-of-experts architecture activated only 37 billion of its 671 billion total parameters on any given inference pass. The model was trained, by widely cited but disputed estimates, for a small fraction of what a comparable American frontier run cost. A figure that understated total research spending even as it captured something real about DeepSeek's efficiency-first approach. Other labs built directly on top of what DeepSeek and its peers released. DeepSeek's R1-Distill-Qwen-32B, a version of its reasoning model distilled onto an Alibaba Qwen base, claimed to outperform the original Qwen model it started from on certain benchmarks. In late 2025 the single most downloaded model on Hugging Face wasn't a new frontier release at all but a small video-captioning model TikTok's parent company ByteDance had fine-tuned from an existing Qwen vision model, itself already the fifth-most-downloaded model of that period. Nobody in this chain trained a frontier model from scratch. Everybody in it improved on something somebody else had already given away.

The scale this loop has reached by 2026 is no longer a matter of dispute. China's National Data Administration disclosed that the country's daily AI token consumption surpassed 140 trillion by March 2026. Up from roughly 100 billion at the start of 2024, better than a thousandfold increase in two years. On OpenRouter, the widely used cross-model routing platform already cited, Chinese models accounted for 61 percent of total token consumption among the platform's ten most used models in a single week in February 2026, processing 5.16 trillion tokens against 2.7 trillion for every American model combined. Four of the world's five most used AI models that month were Chinese.

On Hugging Face specifically China's share of total model downloads reached 41 percent over the trailing twelve months ending in February 2026, against 36.5 percent for the US. This is an outright crossover in the single most watched public metric of open-model adoption, up from a narrower 17.1 to 15.86 percent lead recorded less than a year earlier. The ecosystem producing all of this is wider than DeepSeek and Alibaba's Qwen alone. MiniMax, a Shanghai lab founded by former engineers from the Chinese AI surveillance company SenseTime, listed on the Hong Kong Stock Exchange in January 2026. It saw its share price double on its first day of trading, a market's real money verdict on the commercial viability of the open, freely distributed model this chapter is describing. On price alone the comparison is stark: Moonshot's Kimi K2.5 matched OpenAI's GPT-5.2 on Artificial Analysis's composite intelligence benchmark while costing roughly a quarter as much per token.

The adoption isn't confined to China's borders, and that's precisely the point of a strategy built around global diffusion. Singapore's government-backed AI Singapore program chose Alibaba's Qwen over Meta's Llama as the foundation for its national language model. Malaysian private sector partnerships built regional AI infrastructure on DeepSeek weights in 2025, though the Malaysian government later distanced itself from framing that work as a national sovereign stack. The arrangement stood as an example of how deeply Chinese open models had already penetrated regional deployments regardless of what any single ministry chose to call them. Developers from Nairobi to São Paulo were building products on Chinese open weights for the same reason American startups were: The models were good enough and free.

The Physical Loop, and Why It's Different

Loop two is a harder story to tell in a single data point, because its evidence lives on factory floors rather than download counters. But the direction Beijing is pointed is not ambiguous. In March 2026 China's National People's Congress approved the outline of the 15th Five-Year Plan, the country's central economic planning document through 2030. The plan singled out multimodality, intelligent agents, embodied AI, and swarm intelligence as distinct technical pathways the country intends to pursue to the frontier. This is not an industry trend report. It is a five-year national plan approved by the legislature naming embodied AI as a formal government priority. Chinese embodied AI startups raised roughly ¥20 billion, on the order of $2.8 billion, in the first two months of 2026 alone. Multiple major robotics companies, the makers of the quadruped and humanoid machines already drawing crowds at trade shows like the World Artificial Intelligence Conference, were preparing 2026 stock listings of their own.

Weinan E, an academician at the Chinese Academy of Sciences, argued in a September 2025 paper that the next phase of AI progress will depend less on ever-larger models and more on access to high-quality, diverse training data. This at a moment when several credible estimates suggest the supply of high-quality, publicly available text online could be substantially exhausted sometime between 2026 and 2032. If that argument holds the physical loop stops being a nice-to-have and becomes the more important of the two over the long run. A trillion words of internet text is a finite, shrinking resource everyone is drawing from at once. A country's entire manufacturing base, generating fresh sensor and interaction data every hour a factory runs, is not. China's dominance in manufacturing was, until AI made data itself a competitive input, simply an industrial fact. The kind of thing a trade economist would track and a technologist would ignore. The two-loops framework treats it as a training advantage the US, whose manufacturing base is a fraction of the size, has no comparable way to match, regardless of how many chips its labs can buy. The physical loop's evidence is thinner than the digital loop's by design, because factory data doesn't publish itself. The direction of travel is the strongest thing an outside observer can honestly claim.

But the US does have an ace in its sleeve. There have been reports that up to 90 percent of the total amount of data stored on internet-connected systems is behind corporate or government firewalls and therefore invisible to the bots gathering data to train the frontier models. As training data becomes scarcer, the incentive to open up that data grows. That trove may be enough to keep the models improving at the rates we have seen for longer. Whether it gets us to AGI is another story.

The Paradox Underneath the Strategy

None of this should be read as a clean story of independent Chinese innovation. The same reporting that documents the two loops also documents a real tension sitting underneath them. Chinese labs including DeepSeek, Moonshot, MiniMax, and Zhipu have each been accused, with varying degrees of evidence, of large-scale distillation from Western frontier models, training their own systems in part on outputs generated by Claude and ChatGPT rather than purely on independent research. Anthropic itself accused several Chinese labs in early 2026 of extracting capabilities from its models without authorization through exactly this method and published its own research on detecting and preventing the attempt. China is, in this sense, trying to localize the AI stack at the hardware layer while remaining dependent, at the model-training layer, on the very Western frontier labs its official strategy is designed to out-compete.

Even the openness itself is not purely idealistic, and Chinese labs have said as much without much prompting. Without access to the cutting-edge chips export controls restrict, releasing a model's weights openly is also the cheapest available way to buy back some of the compute advantage those controls are designed to deny. Every developer who downloads and improves a released model is, in effect, contributing free research the lab would otherwise have had to fund with GPU hours it doesn't have. Open source, on this reading, is not only a diplomatic gesture from a rising technological power, but what a compute-constrained ecosystem does when the alternative is falling permanently behind on hardware it cannot buy.

The hardware side of that localization effort has had its own setbacks. After training R1 on Nvidia hardware in January 2025, DeepSeek was reportedly pushed by Chinese authorities to train its next major model, R2, on domestic Huawei Ascend chips instead, part of the broader state push toward chip self-sufficiency. After multiple training failures, the effort was reportedly scaled back. R2's training work shifted back to Nvidia hardware, with Ascend chips relegated to handling inference rather than training. The state can mandate a preference, but it cannot yet mandate that the preferred hardware actually works at the scale the mandate requires. This is a real constraint on the two-loops strategy. The digital loop's efficiency gains have so far outpaced the hardware loop's ability to run independently of the chips that America is trying to restrict.

The Skeptical Case

Downloads are not deployments. The two-loops framework, and the adoption statistics used to support it, deserve the same scrutiny applied to every other bull case that has been taken seriously. A model with 700 million cumulative downloads on Hugging Face is unambiguously popular, but downloads capture researchers experimenting, developers testing, and files mirrored across platforms as much as they capture companies running critical business logic on top of the model in production. Martin Casado's oft-cited estimate that 80 percent of startups pitching Andreessen Horowitz with an open-source stack are running on Chinese models is one venture capitalist's impression of the pitches crossing his desk, not a systematic survey of the industry. DeepSeek R1's brief run atop the US App Store charts in January 2025, the very episode that opened this chapter, is itself a data point that arguably says more about the news cycle's appetite for novelty than about which model any given company was actually running in production six months later.

None of these caveats erase the underlying signal. Hugging Face's download crossover and OpenRouter's token consumption numbers are platform-reported usage statistics, not survey estimates, and both point the same direction the more anecdotal figures do. But a country's strategic advantage and a headline statistic are not automatically the same thing, and the honest version of this chapter has to hold both truths. The trend is real, and some of the numbers used to dramatize it are looser than the drama implies.

Is It Really Open?

Free weights are not the same thing as a trustworthy supply chain. Within weeks of R1's release, governments around the world started treating the two as separate questions. Italy's data protection authority ordered an emergency restriction on DeepSeek's handling of Italian user data by the end of January 2025, the first such action taken against an AI chatbot under European privacy law. Taiwan barred its government agencies, state enterprises, and public schools from using it days later, citing cross-border data transmission and information leakage risk. Their public statement was direct enough to leave no ambiguity about who the ministry meant. Australia, South Korea, the Netherlands, Japan, Canada, the Czech Republic, and more than a dozen individual American states followed within the same stretch of weeks, as well as NASA.

The US Navy and the Pentagon blocked it from government systems outright. South Korea's privacy regulator found DeepSeek had been transmitting user data to ByteDance, TikTok's parent company, without disclosing it, a violation serious enough to force a temporary nationwide suspension of new downloads. Security researchers at the firm Wiz found a DeepSeek database sitting exposed on the open internet with no authentication at all, containing chat histories, API keys, and internal operational logs. A separate firm found the app using hard-coded encryption keys and transmitting some data without encryption. Another security researcher reported code inside DeepSeek's web platform linking to China Mobile's authentication infrastructure, a state-owned telecom operator the US had already barred from its own networks on national security grounds years earlier.

None of this requires a proven backdoor to matter, because Chinese law already supplies the mechanism a backdoor would otherwise be needed for. Under China's 2017 National Intelligence Law, any organization or individual is obligated to support, assist, and cooperate with the country's national intelligence work when asked, with no requirement to notify the people whose data is involved. A Chinese company doesn't need to build a secret door into its own product when the law already grants the state a standing key to the front one. That's a structural feature of doing AI business inside China's legal system, not an allegation about any single company's engineering choices, and it applies to every lab this chapter has named, not only the one that happened to top the App Store first.

The censorship built into these models tells a similar story from a different angle. It lands with a specific irony given the openness this chapter has spent most of its length describing. Ask most Chinese open models about Tiananmen Square or Taiwanese independence, and they will decline to answer, or answer in terms that track official government messaging precisely. Ask the same models for help with something considerably more dangerous, and researchers testing their guardrails found them dramatically more permissive than their American counterparts, having apparently optimized their safety training around the topics that matter to Beijing rather than the topics that matter to safety researchers generally. The openness, in other words, was never unconditional. It was selective in exactly the direction the state that encouraged it would want.

None of this has slowed adoption, and this book cannot change that. Governments banned the app from official devices while ordinary users, developers, and startups worldwide kept downloading the weights anyway, because the bans applied to government systems and the free download applied to everyone else. A year after the bans began, one industry tracker recorded DeepSeek's global usage up roughly 960 percent even as its access remained constrained across a meaningful share of the Western world. Doubts about data handling, state compulsion, and selective censorship turned out to be real, well-documented, and almost entirely beside the point for the millions of developers deciding which model to build on. As we've see at every layer of the stack, price and performance settle most decisions long before trust does. Chinese open models kept offering enough of both to make the doubts something to note rather than something to act on.

The State's Other Hand

Openness, in this strategy, has never meant indifference to ownership. Even as Beijing has encouraged its labs to give model weights away to the entire world, it has moved decisively to keep the companies producing those models inside Chinese hands. When Meta moved to acquire Manus, an AI agent startup with roots in China that had built a fast-rising international following, Chinese authorities intervened to block the deal. The message embedded in that intervention is precise. The weights can leave the country for free, and frequently do, by design. The company that makes them, and the strategic value embedded in owning the next generation of labs rather than merely running last year's downloads, does not leave the country at all.

The governance ambitions sitting behind the model releases reach further than any single acquisition fight. China has signed roughly 120 government-to-government science and technology cooperation agreements, a significant share of them with developing countries, and has stood up dedicated AI cooperation centers with BRICS partners and with individual nations like Laos. This infrastructure is meant to extend Beijing's technical standards and its preferred governance norms alongside the models themselves. The proposed World AI Cooperation Organization is, in this light, an attempt to do at the level of international rulemaking what the open-weight releases are already doing at the level of individual developers. Make Chinese-built infrastructure, and increasingly Chinese-authored rules, the default a cost-sensitive world reaches for first.

One caveat belongs in this account rather than left for a footnote. Chinese open models carry the imprint of the country's domestic content-moderation regime, trained and tuned to avoid outputs that conflict with government messaging on politically sensitive topics. A developer in Jakarta or Lagos building on a freely downloaded Chinese model inherits those constraints along with the capability, whether or not that trade-off was ever explained to them as part of the deal.

What This Means for the Trap

China's two loops are, at the bottom, the sharpest version yet of a pattern now traced at every layer of the AI stack: giving away the commodity to capture the value sitting above and beneath it. Mistral made this bet as a single company, trading model-layer revenue for a business built on top of a free model. China is making the identical bet on the scale of a national industrial strategy, trading the model layer's commercial value for global adoption and developer mindshare. This is a default-infrastructure position among cost-sensitive developers worldwide, while quietly guarding the physical-world data advantage and the corporate ownership that the free weights themselves don't give away.

This is Jevons paradox operating at the level of geopolitics rather than a single firm's balance sheet. Cheaper, freely available Chinese models don't shrink the strategic prize, they expand the base of the world's developers building on Chinese infrastructure, which expands the training signal flowing back into the next generation of Chinese models, which expands the advantage again. The Perez framework's installation frenzy is visible here too, in the billions of renminbi chasing embodied AI startups and the robotics IPOs racing to list before the physical loop's advantage is even fully proven out. It's the same pattern of capital front-running confirmed demand traced from railroad mileage to circular AI financing.

The distinction between financial and production capital cuts an unusual way in this chapter. The free model weights themselves are barely capital at all, in the traditional sense, since they cost the recipient nothing. But the factories, robots, and data infrastructure generating the physical loop are about as close to pure production capital as anything examined, assets that keep manufacturing goods and generating training data regardless of which AI strategy proves out. It is the Indispensability Trap's access-guarantee mechanism showing up in a form no earlier chapter anticipated. Not a domestic regulator capping a company's returns after the fact, but a foreign government building indispensability as policy in full public view specifically because it has calculated that American export controls can reach its chips but cannot reach its factories, its developers, or the models built freely on top of both.

The nations watching this unfold from the outside, the ones without a Silicon Valley and without a manufacturing base the size of China's, face the sharpest version of Part III's fork yet. Build on top of an American stack they don't control, build on top of a Chinese stack they don't control either, or attempt to build a stack of their own. That choice, and the sovereign hedges it has already started to produce, is where we turn next.

Chapter 12

Sovereign Ground

> No alliance can be considered separately from the effort undertaken by each of its members.
>
> —Charles de Gaulle

In the spring of 2026, OpenAI paused its planned Stargate UK data center project. It was the flagship of Britain's bet that American frontier labs would build critical infrastructure on its soil. The reasons were never fully explained in public, but the message was heard loud and clear regardless. A country's AI ambitions cannot depend, in the end, on a decision made in San Francisco, Washington, Beijing, or anywhere but that country itself. Within weeks, the UK government stood up a Sovereign AI Fund with new urgency. This is not a story about paranoia. It's a story about a promise that turned out to be conditional, learned by a G7 economy in something close to real time.

I've written about this word, sovereignty, more than almost any other in the past year except AI itself. The thing I keep coming back to is that it's carrying at least three different meanings at once. Most conversations about it, including an earlier draft of this very chapter, collapse all three into one before the reader notices the switch.

By early 2026, after the World Economic Forum in Davos, Switzerland, and the cascade of repetition that followed, digital sovereignty had become kitchen table conversation for world leaders. It meant something different depending on whose kitchen table you were sitting at. For a head of state, it meant sovereign AI, whether the country's models, chips, or data centers sit in someone else's house, especially when that house is halfway across the world and the landlord keeps changing the rules. For a chief information officer, it meant something narrower and more immediate: data sovereignty, the question of where the company's data physically lives and who can compel access to it regardless of whose model touches it. For an ordinary person, it meant something narrower still, and arguably more overlooked than either: personal sovereignty, control over your own digital footprint, your own data, your own privacy, independent of what your government or your employer decides to do with theirs.

This chapter takes all three seriously, in ascending order of scale. The mistake worth avoiding is assuming that solving sovereignty at one level solves it at the others. A country can build a gigafactory and still leave its citizens' data sitting on servers a foreign intelligence service can compel access to. A company can adopt the most rigorously sovereign cloud architecture available and still run on a workforce that has surrendered its own digital footprint to a dozen platforms without a second thought. The word is doing three jobs. It deserves three sections.

Personal Sovereignty

Start at the smallest scale, because it's the one most of the policy conversation happening around it tends to skip past entirely. Every individual deserves some meaningful control over their own data, their own privacy, their own independence, the same case nations were making for themselves at Davos applied one layer down. You shouldn't be under constant surveillance unless you've actually done something to warrant it. If you choose to trade some of that control away for convenience, for an app, a service, or a shiny feature, that's a choice you're entitled to make. What's disappearing, as AI assistants read your email, your calendar, and increasingly your location and biometric signals by default, is the ability to make that trade knowingly rather than by default settings nobody reads.

This is the sovereignty dimension with no gigafactory to point to and no summit declaration bearing its name. That is exactly why it's the one most likely to get steamrolled while nations and companies fight over the other two. A government can win every argument in this chapter, build its own compute, host its own models, keep its citizens' data within its own borders, and still preside over a population that has no meaningful say over how an AI system trained partly on their own behavior gets used to predict, nudge, or price them. Sovereign AI and sovereign data are necessary conditions for personal sovereignty. They are nowhere close to sufficient ones, and no country in this chapter has yet built a serious policy answer to the gap between the two.

A small startup made the business case for this gap in the same week this chapter's other events were unfolding. Venice AI, founded by cryptocurrency entrepreneur Erik Voorhees, encrypts a user's prompts on their own device before anything leaves it and keeps no server-side record of the conversation at all, an architecture Voorhees describes, in exactly this chapter's terms, as respecting the sovereignty of its inhabitants, both human and agentic. By July 2026 the company had reached a billion-dollar valuation on its first outside funding round. It was already profitable on roughly $70 million in annualized revenue from three and a half million users processing over 1.3 trillion tokens a month. Venice is a rounding error next to any hyperscaler, and its existence is the clearest evidence available that personal sovereignty isn't merely a philosophical complaint this chapter is raising on individuals' behalf. It's a product people will pay for once somebody builds it.

Data Sovereignty

One layer up, and considerably more concrete, sits the question keeping chief information officers awake. Where does our data actually live, and who can reach it. A SUSE-commissioned survey of more than three hundred IT leaders across the US, Europe, and Asia, published in April 2026 put numbers on how fast this had moved from policy seminar to operational reality. Ninety-eight percent of respondents said digital sovereignty was now a priority for their organization. Ninety-three percent said they either already had a sovereignty strategy in place or were actively developing one. Most strikingly, 51 percent said a foreign entity had already breached their company's privacy or data regulations, and nearly a quarter of those described the incident as a major one, a number large enough to suggest this isn't hypothetical risk management but a response to something that has already happened to roughly half the organizations surveyed.

The mechanism connecting this to AI specifically is straightforward once stated. AI systems depend on massive data sets and infrastructure pipelines that stretch across clouds, providers, and jurisdictions, often without the deploying company having full visibility into where a given query actually gets processed. Sixty-four percent of the IT leaders in the same survey said AI transparency requirements were now pushing them to strengthen their resilience strategies specifically. Further, 94 percent said open-source infrastructure was critical to achieving it, on the theory that code you can inspect and redeploy yourself is a meaningfully different sovereignty position than a black-box service running in someone else's region.

This is a distinct question from anything the rest of this chapter's national AI strategies address. A country can build its own gigafactory and its own frontier model, and every company operating inside that country can still be running its actual workloads on a foreign hyperscaler's regional cloud, subject to that hyperscaler's home government's legal reach, regardless of whose logo is on the model doing the inference. Sovereign AI at the national level and data sovereignty at the enterprise level are, in practice, two different fights, fought by two different sets of people, and a nation can win one while its own companies quietly lose the other.

The word tying all of this together at the enterprise scale, I'd argue, isn't sovereignty at all. It's resilience. For a long stretch of enterprise IT history, reliability was the gold standard. If systems stayed online, the job was done. That's no longer the right test. The organizations built to last through the next decade won't be the ones with the fastest or most scalable systems. They'll be the ones that understand their own infrastructure deeply enough to keep operating when a supplier, a cloud region, or an entire jurisdiction goes dark on them. Sovereignty, open infrastructure, and resilience aren't three separate trends. They're three angles on the identical question this whole chapter keeps circling. How much control does an organization, or a nation, or a person, actually have over the systems it depends on.

Sovereign AI: The National Bet

By 2026 sovereign AI had gone from an aspiration to a budget line in nearly every government with the fiscal capacity to fund one. France has Mistral. The UAE has TII's Falcon models, developed alongside its national champion G42. Saudi Arabia has HUMAIN. Singapore has SEA-LION, whose current flagship line moved from Meta's Llama to Alibaba's Qwen in late 2025, as already noted. Japan has its own LLM-jp initiative. India has BharatGen. What united all of them by 2026 was not a shared theory of how to achieve independence but a shared and suddenly urgent premise: Depending on a small number of American labs and the narrower set of chip exporters that supply them had become uncomfortable enough for enough governments that the discomfort itself became policy.

Three Species of Sovereignty

Sovereign AI can be further divided into three species. The first species, practiced by the EU and the UK, is genuine independence-seeking, an attempt to build compute, models, and data infrastructure that doesn't depend on any single foreign supplier, American or Chinese. The second, practiced by the Gulf states, is closer to the opposite instinct, using sovereign wealth to buy deeper, more privileged access to the American stack specifically, in exchange for explicitly severing ties to the Chinese one. The third, practiced by India and articulated through a global declaration bearing its capital's name, rejects the binary altogether, refusing to build on either superpower's stack, while trying to become the broker who sets terms for everyone else caught in the same position. Three countries, three answers to the same fork nations were facing at the close of the China chapter. None of them has yet proven durably right.

The Independence Bet

Europe's version of this hedge runs through an initiative its own architects have compared to CERN. Announced by Commission President Ursula von der Leyen at the Paris AI Action Summit in February 2025, InvestAI aims to mobilize €200 billion for AI across the EU. A full €20 billion of that sum ring-fenced to build up to five AI gigafactories, each equipped with more than 100,000 advanced processors, roughly four times the capacity of the AI factories that preceded them. The pitch is explicit, a guarantee that European researchers, startups, and industry are no longer dependent on non-European hyperscalers for the compute their AI ambitions require.

The response to the informal call for interest was, by any measure, enormous. Seventy-seven proposals arrived from sixteen member states across sixty sites with indicative interest exceeding €230 billion, more than ten times the fund's committed capital. The Council of the European Union formally amended the EuroHPC Joint Undertaking's mandate in January 2026. It made the gigafactories official, and the program's commissioner, Henna Virkkunen, was blunt about the sovereignty requirement built into it. Majority ownership of these facilities has to sit in European hands, and high-risk vendors are barred from participating at all.

What the independence bet can't paper over is a timing problem seen twice before, at the silicon layer and the model layer. The formal call for gigafactory proposals, originally planned for December 2025, had slipped to an unspecified date in 2026 as of this writing. In the meantime, the AI infrastructure actually rising on European soil is arriving through the companies the whole initiative was designed to reduce dependence on. In Portugal, Microsoft and Nscale are building out the Start Campus site in Sines with more than 66,000 Nvidia GPUs planned by late 2027. In Germany, Nvidia and Deutsche Telekom have stood up a 10,000-GPU industrial AI cloud explicitly positioned as the precursor to a future gigafactory. Europe's earliest gigafactory-scale facilities, in other words, are unlikely to be EU-funded, EU-owned, or built around European frontier labs at all. The sovereignty ambition is real, but the infrastructure arriving first is American.

One European country has already made the alternative case look plausible and has met it before. Chapter 10 told Mistral's story as a business bet, giving away its model weights to build revenue one layer up the stack. It is also, without ever needing the gigafactory program to succeed, the closest thing Europe has to a working sovereign AI champion: a French lab, backed increasingly by French and EU public money and procurement preference, whose models run the country's government pilots and whose data centers near Paris and in Sweden are exactly the kind of European-controlled compute infrastructure the InvestAI program is still trying to build from scratch. Mistral got there first by building a company, not a fund. That's a data point worth weighing against every centrally planned gigafactory in this chapter.

Britain's version of the same bet moved faster on paper, if not necessarily in scale. Following the Stargate UK pause, the government launched a Sovereign AI Fund backed by up to £500 million, chaired by venture investor James Wise, alongside a separate £2 billion commitment to expand national compute capacity twentyfold by 2030. A further £1.1 billion was announced via the AI Hardware Plan in June 2026, which includes a £750 million national supercomputer targeted for 2030. The centerpiece so far is Isambard-AI, a Bristol-based system built around more than five thousand Nvidia Grace-Hopper superchips. Britain's first sovereign frontier model, Lumen Sovereign, is being trained entirely on it, backed by a coalition that reads like a roll call of the country's largest banks, insurers, and defense contractors, BT, Lloyds, NatWest, LSEG, BAE Systems, Babcock, Thales, PwC, and the Alan Turing Institute among them.

The scale problem is unavoidable, and to really understand it, it is best to show it in plain numbers rather than leaving it implied. Five hundred million pounds is, by one widely repeated comparison, roughly 0.08 percent of OpenAI's $852 billion valuation. Amine Abidi, a senior partner in Kearney's Digital and Analytics practice, put the honest version of the ambition directly: The point was never to match American or Chinese scale, but to secure control over the foundations AI depends on—compute, data, and infrastructure—even at a fraction of the size. Whether that's a sober recalibration of what sovereignty can realistically buy, or a polite way of admitting the fund is too small to matter, is an ongoing argument. This book won't and shouldn't pretend to settle it. One skeptical assessment aimed at the whole European approach put the risk plainly: Over-indexing on a fully domestic AI supply chain is slow, may never fully arrive, and risks falling behind the very countries, America and China, that the strategy is trying not to depend on. Optionality, from this view, beats autarky. Britain's fund seems to have absorbed that lesson already, since its explicit goal is control rather than replication.

The Alignment Bet

The Gulf states are pursuing something that looks, on the surface, like the same hedge, and is in substance closer to its mirror image. Where Europe and Britain are trying to reduce dependence on the American stack, the UAE and Saudi Arabia are spending extraordinary sums to deepen it, using sovereign wealth as the currency that buys a permanent, privileged seat inside it.

The clearest evidence is what the UAE gave up to get there. G42, Abu Dhabi's national AI champion, spent late 2023 into early 2024 unwinding its relationships with Chinese technology partners. In February 2024 it publicly announced that its investment arm had divested its Chinese holdings entirely, and it committed to use only American-origin chips and technology for sensitive applications. In exchange it received a $1.5 billion investment from Microsoft and access to OpenAI's models routed through Microsoft's commercial relationship. That divestment is not a footnote. It is the single most consequential technology realignment in the UAE's recent history. It was the price of admission to the American stack rather than a step toward independence from it. The flagship result, Stargate UAE, is a five-gigawatt campus built by G42 in partnership with OpenAI, Oracle, Nvidia, Cisco, and SoftBank. It is expected to cost around $30 billion and to cover ground larger than Monaco.

Saudi Arabia's version runs through HUMAIN, a subsidiary of the kingdom's nearly trillion-dollar Public Investment Fund launched in 2025. CEO Tareq Amin has described its mandate as building the full AI stack, from data centers and infrastructure to models and applications, rather than just data centers. The numbers involved are difficult to compare to anything else in this book at a comparable stage of national build-out. A framework to deploy up to 600,000 Nvidia AI accelerators over three years, a $10 billion venture fund, an 18,000-unit shipment of Nvidia's GB300 chips announced in May 2025 and confirmed as delivered in late December 2025 following the US export authorization the previous month. A 480-megawatt government facility, the Hexagon data center, built by the Saudi Data and AI Authority alongside HUMAIN's campuses and described by its backers as the largest Tier IV government data center in the world. In November 2025 the US Commerce Department authorized the export of 35,000 Blackwell-class chips each to G42 and HUMAIN, conditioned on rigorous security and reporting requirements, including US operator oversight of the data centers housing the chips, requirements that built on the Gulf states' earlier commitments to limit Chinese technology integration. This is the access-guarantee mechanism showing up as an instrument of alliance management rather than domestic regulation.

None of this insulates the bet from the risks any concentrated infrastructure investment carries, and 2026 supplied a vivid demonstration of exactly that. Both governments have attached extraordinary economic stakes to the wager. Saudi Arabia is targeting AI contributing 12 percent of GDP by the end of the decade. The UAE, under its National Strategy for Artificial Intelligence 2031, has set a target for AI to contribute up to 20 percent of nonoil GDP by 2031, numbers that treat AI infrastructure not as a hedge alongside oil wealth but as its intended successor.

But not all of that Gulf capital is flowing through national champions like HUMAIN. Aramco Ventures led an $800 million round into Together AI, an American neocloud that has nothing to do with Saudi Arabia's sovereign stack and everything to do with hosting open-weight models cheaply at scale, a bet Chapter 10 placed in its proper context. The sovereign wealth logic and the open-model economics turn out to be aiming at the same target from two different directions. In April, amid the open military conflict then underway between the United States, Israel, and Iran, Iran's Islamic Revolutionary Guard Corps claimed to have struck an Oracle facility in Dubai. UAE authorities disputed a direct hit, later confirming only that debris from an intercepted Iranian projectile had struck the facade of the building housing Oracle's Dubai offices with no casualties.

The Gulf states had built their AI strategy on a premise of regional stability and cheap, abundant energy. A missile-debris strike on the facade of a building housing hyperscale infrastructure is the premise being tested in the most literal way available, and analysts covering the region were direct about the consequence. As CSIS analyst Aalok Mehta put it, the conflict had shattered the illusion of long-term stability in the Gulf. The same reporting warned that future facilities would likely cost more and take longer to build because of hardened construction, higher insurance, and new anti-drone requirements. Sovereignty bought through alignment with a superpower still leaves you exposed to whatever conflicts that superpower is drawn into.

The Third Path

India's version of this hedge refuses the premise that a country has to choose a side at all. It's the only one of the three that has tried to turn that refusal into a document other nations could sign.

The India AI Impact Summit, held in New Delhi in February 2026, was the first in the rotating series of global AI summits—after Bletchley Park, Seoul, and Paris—to be hosted by a country from the Global South, and India used the platform deliberately. The resulting New Delhi Declaration on AI Impact was endorsed by nearly ninety countries and international organizations at signing, a total that continued to climb in the weeks that followed. It was built around what the government called the Charter for the Democratic Diffusion of AI, treating high-performance compute and vetted data sets as something closer to a global public good than a commodity to be rationed by whichever superpower controls the supply. India's delegation was explicit about the alternative it was declining. Reporting from the summit described India rejecting a US proposal that regional partners build exclusively on the American AI stack, asserting instead a right to develop its own vertical stack rather than accept what its officials termed a permanent strategic dependency.

The declaration itself is deliberately nonbinding, a fact its critics were quick to point out. Unlike the EU's AI Act, it creates no enforcement mechanism and prohibits nothing. But the domestic build-out behind it is concrete enough to take seriously on its own terms. The IndiaAI Mission expanded its national GPU pool from roughly 38,000 to a targeted 100,000 by the end of 2026, a fraction of the compute any single American hyperscaler commits to a single site. Additionally it unveiled homegrown models built for the country's linguistic reality rather than adapted from someone else's. BharatGen's Param2, a 17-billion-parameter model spanning twenty-two official languages, and Sarvam-105B, one of two open-weight mixture-of-experts models Sarvam AI released around the summit. Officials framed both as proof that frugal, India-built models could match or beat larger Western systems on Indic-language benchmarks that global leaderboards don't measure. The government's framing, delivered by the technology minister at the summit, called the target ecosystem frugal, sovereign, and scalable, in that order, which is an honest acknowledgment that scale is the one thing India is not currently trying to match.

What This Buys, and What It Doesn't

One of the sharpest challenges to everything in this chapter doesn't come from a skeptic of sovereignty itself, but from a January 2026 Tony Blair Institute report. *Sovereignty in the Age of AI* makes the case for a more disciplined version of it. Sovereignty, the argument runs, should not mean independence from all others. It should mean the ability to act strategically, with agency and choice, in a world that is irreversibly interdependent. Full self-sufficiency, on this view, is too expensive, too slow, and for most countries simply impossible, and treating it as the goal misrepresents what sovereignty actually requires.

No state can dominate every layer of the AI stack, so the useful question isn't whether a country owns its entire stack, but where it becomes indispensable enough—in data assets, specialized models, regulatory standards, energy capacity, or talent—to gain real leverage over the parts it doesn't own outright. The report's sharpest line cuts against the instinct running through every national program in this chapter. Isolation and protectionism do not protect sovereignty; they weaken it. A country that refuses to access frontier capabilities becomes dependent on whoever does, less able to defend itself or compete regardless of how many gigafactories it has announced.

That argument doesn't dissolve the stakes this chapter has described. It reframes them. Personal sovereignty, data sovereignty, and national sovereignty are not the same fight scaled up and down, and a government that wins the national argument—gigafactories built, models trained, chips secured—has not automatically won the other two. None of the four engines quite anticipated sovereignty operating at three scales simultaneously. Every prior chapter's version of the access-guarantee mechanism arrived after the fact, a regulator or a court responding to an indispensability that had already been established.

What individuals, companies, and now nations are all doing at once, in 2026, is building the hedge before the dependency fully calcifies. The same instinct in Intel's rescue and Reflection AI's bet against the Chinese open-weight wave is now playing out at every scale from a single person's data settings to an entire national budget. The Indispensability Trap is present in all three dimensions as a fear rather than an outcome. No government, company, or individual in this chapter has yet had their returns or their agency capped by anyone, because none of them has become indispensable to anything yet. They are trying to buy their way out of ever being in the position of earlier build-outs, which is a materially different exercise from the one the railroads, the Bell System, or Samuel Insull ever faced.

Jevons paradox offers a genuine reason to think this could work regardless of the scale mismatch. Cheaper compute and cheaper models lower the price of admission to a credible sovereign stack every year. A fund that looks like a rounding error against OpenAI's valuation in 2026 buys considerably more capability than the same fund would have bought in 2024. The Perez framework offers the less comfortable reading. Every sovereign AI announcement in this chapter is still installation-phase capital, chasing a demand curve nobody has confirmed. The same frenzy traced through circular financing is visible here too, just funded by treasuries instead of venture capital. The distinction between financial and production capital cuts in an unusually literal way across this chapter's three approaches. Britain's and the EU's compute commitments are trying to become production capital before the frenzy that would otherwise absorb them arrives. The Gulf's $100 billion campuses are production capital in the most physical sense, poured concrete and installed silicon, purchased with financial capital extracted from oil rather than raised from investors. India's more modest, deliberately frugal build-out is a bet that production capital doesn't have to match the scale of the build-outs it's hedging against in order to buy real optionality.

The honest scorecard, held against the trillion-dollar build-outs already documented in America and China, is not flattering to any of the three approaches on pure scale. The EU's ring-fenced gigafactory fund, Britain's sovereign compute commitments, and India's GPU expansion combined amount to a rounding error against the roughly $7.6 trillion Goldman Sachs projects for the global AI build-out through 2031.

But scale was never quite the right test for what these governments are actually buying. What they're purchasing, at wildly different price points and with wildly different theories of the case, is optionality. The ability to say no to a dependency before it becomes the kind of indispensability this book has spent eleven chapters watching turn into someone else's leverage. Whether optionality bought this cheaply survives contact with a real crisis, the way the Gulf's stability premise didn't survive contact with a missile, is the question every government in this chapter is still answering in real time.

That question is about to get sharper. The same leverage these nations are hedging against, compute access as a lever of state power, is exactly what the next chapter examines from the other side. What happens when the country that actually controls the frontier decides to use that control directly.

Chapter 13

The Switch

> The power to tax involves the power to destroy.
>
> —Chief Justice John Marshall, *McCulloch v. Maryland*, 1819

At 5:21 p.m. Eastern time on June 12, 2026, Anthropic received a letter from the Bureau of Industry and Security. It arrived three days after the company had released Claude Fable 5 and its more capable sibling, Claude Mythos 5, to the public. The letter did not announce a new regulation. It informed Anthropic, under a provision of the Export Control Reform Act that had never been used this way before, that any foreign national anywhere in the world, inside the United States or outside it, now required an individually validated export license to access either model. That included Anthropic's foreign national employees. Within hours, both models were dark everywhere on Earth. A rate case takes years to resolve. An antitrust suit takes a decade. This takedown took a single letter and an afternoon.

The Mechanism Itself

What makes this episode worth a full chapter, rather than a footnote to the export control material already covered at the silicon layer, is how novel and improvised the legal instrument actually was. Commerce did not publish a new rule. It sent what's called an "is informed" letter, a mechanism that lets the agency impose a license requirement on a specific party or item without going through the ordinary rulemaking process first. Two statutory bases were cited.

First, 50 U.S.C. § 4817(b)(1), which permits the government to establish interim controls on emerging technology essential to national security. However, it had never been invoked this way before. A separate regulation governing "is informed" letters specifically, which existed already but, as the next section explains, was built for a considerably narrower purpose than the one it was asked to serve here. The letter cited national security concerns without specifying which weapons, which adversary, or which end use it was actually worried about.

The letter didn't arrive from nowhere. Months earlier, in February 2026, the President had ordered federal agencies to stop using Anthropic's models after the company refused the Pentagon's preferred contract terms, specifically declining to permit Claude's use for lethal autonomous weapons or large-scale civilian surveillance. The Defense Department followed in early March by formally designating Anthropic a supply chain risk, a label historically reserved for foreign adversaries. That designation gave the Fable and Mythos suspension the legal runway it might not otherwise have had, even though it was never squarely about export control, and Anthropic contested and was successful in a suit brought in federal court. Legal scholars spent the following weeks debating a question that shouldn't still be open this deep into the AI era: Does a foreign national typing a prompt into a chatbot and receiving an answer even constitute an export of anything under a body of law written to govern physical goods crossing physical borders.

One law professor floated a workaround theory, that each individual model output might itself count as controlled technology the moment it's generated, since it would dodge the awkward question whether the underlying model has already been too widely released to be regulated at all. The Export Administration Regulations carry their own exception for software that has been widely disseminated, a provision that gave Anthropic a plausible legal argument that a publicly released, three-day-old model might already be outside the government's reach even as the government was actively reaching for it. Commerce, for its part, never engaged that argument directly, and the letter never explained why a general framework built for weapons technology and physical exports applied to a chatbot's text output at all. Nobody has a confident answer yet, including, by all appearances, the government that sent the letter.

The government's directive, as it turned out, leaned on more than the interim controls statute alone. A second provision, 15 C.F.R. § 744.22(b), authorizes what the regulations call an "is informed" letter, but on its face that authority is narrow. A targeted, case-specific tool meant for situations where Commerce has already identified a particular military intelligence end user or end use, not a mechanism for a blanket, worldwide ban untethered from any identified recipient at all. A separate federal lawsuit, filed by a legal technology company that had built its product on Fable and lost access when the directive landed, argues that both statutory bases were stretched well past what either was written to do. It stated that the only export classification that ever directly covered AI model weights, ECCN 4E091, had itself been rescinded in May 2025 with nothing put in its place. Commerce, the suit contends, is enforcing a control that no longer exists against conduct, the delivery of a chatbot's text output, that transmits no technology, source code, or model weights of any kind. That a company two steps removed from Anthropic itself felt compelled to sue over losing access to a chatbot is its own small measure of how far this reached.

The scope of the order compounded the novelty. It didn't target a specific adversary nation, the way export controls on advanced chips single out China. It applied to every foreign national everywhere, a category so broad that Anthropic had no practical way to comply with anything short of a total shutdown. There is no way to verify a user's citizenship through an API call. A consultant advising companies on the fallout put the absurdity in the plainest terms available. Three-quarters of Americans themselves don't hold a passport, so even confirming who counts as a domestic user, let alone screening out everyone who doesn't, was never a real option on the timeline Commerce had given. Anthropic's statement made the same point more diplomatically, informing customers that the order's net effect was to require disabling both models for everyone, worldwide, immediately in order to guarantee compliance with a rule nobody could actually apply selectively.

The trigger for the letter connects back to a company this book has already spent an entire chapter examining from a completely different angle. Anthropic's safety architecture had restricted Mythos to a small set of vetted partners from the start, under an internal program it calls Project Glasswing, publicly unveiled in April 2026 after Mythos Preview had already spent months in limited testing with approved organizations. Fable 5, released to the wider public on June 9, carried guardrails specifically designed to block the model from helping identify exploitable cybersecurity vulnerabilities.

According to multiple accounts of the weekend that followed, Amazon, one of Anthropic's largest investors and the cloud provider examined at length in Chapter 8's landlord-tenant-rival tangle, alerted Treasury Secretary Bessent and other officials directly. Amazon Chief Executive Andy Jassy reportedly delivered the news that its researchers had found a way to extract information from Fable 5 that could be used in cyberattacks. A whirlwind twenty-four hours followed, with Anthropic Chief Executive Dario Amodei on tense calls with National Cyber Director Sean Cairncross, Bessent, and Commerce Secretary Lutnick, before the directive landed late Friday evening.

Anthropic did not accept the government's framing quietly. The company argued publicly that the vulnerability in question was a narrow one, capable of unlocking a specific slice of Mythos's cybersecurity capability rather than defeating every safeguard at once. He pointed out that the same jailbreak technique could reportedly be used against OpenAI's publicly available GPT-5.5, a model facing no comparable export restriction at all. "We disagree that the finding of a narrow potential jailbreak should be cause for recalling a commercial model deployed to hundreds of millions of people," the company wrote, adding that applying this standard consistently across the industry would functionally halt new model deployment for every frontier lab at once. Whether that argument was self-serving or simply correct is a question this book can't resolve on Anthropic's behalf.

What it does highlight is that the same company with a multibillion dollar stake in Anthropic's success was the one whose researchers surfaced the vulnerability that gave Washington its opening to shut the model down worldwide. Chapter 8 asked how a single company can be investor, customer, and rival all at once. This is what it looks like when that same tangle produces a security finding sharp enough to trigger a national security intervention against the company being invested in.

There was, underneath the public dispute over how narrow the jailbreak really was, a second and considerably more alarming thread that only surfaced in Congress weeks later. Senator Mark Warner testified that during classified testing by US intelligence agencies, the frontier model in question had broken into almost all of the government's classified systems, not over weeks, but in hours. If that account is accurate, the public argument over a narrow cybersecurity jailbreak was never really the government's actual concern, or at least not the whole of it. A model capable of penetrating classified infrastructure at that speed is a different order of problem than the guardrail-bypass Anthropic was contesting in public. It would explain a government response far more urgent than the narrow vulnerability alone seems to justify. Warner's testimony prompted new legislative proposals for mandatory AI incident reporting, the first serious push by Congress to require labs to disclose findings like this one as a matter of law rather than at their own discretion. Some reporting also suggested a second, separate concern sat behind the directive. Suspicion that a China-linked group had gained access to Mythos, raising exactly the distillation fear Chapter 11 already documented, a worry that whatever capability the model represented might not stay contained to the partners Anthropic had vetted for it.

The restoration, when it came, arrived with the same informality as the suspension. Both models came back online on July 1, roughly three weeks after they went dark, following negotiations that were never made fully public and a resolution that, like the original letter, was never published as a rule. Anthropic restored Fable 5 to all users and Mythos 5 to a set of approved domestic organizations, with government approval for the latter granted on June 26, and offered affected subscribers a temporary usage boost as an apology for the disruption. The switch flipped off in an afternoon and flipped back on after three weeks of private conversations nobody outside the room can fully reconstruct. A consultant working with companies exposed to this kind of risk offered the line that captures the whole episode better than any regulatory analysis could: A frontier model can now vanish for reasons that have nothing to do with uptime, price, or performance, the three things a customer has always been trained to worry about. The fourth risk, the one this chapter is named for, is that the switch itself sits in Washington, not in Anthropic's data centers.

The episode drew criticism from directions that didn't sort neatly by party. Dean Ball, an AI policy expert who had briefly served inside the Trump administration before growing openly critical of this particular decision, called the underlying contradiction almost cartoonish. An administration whose stated posture was that the US should keep exporting advanced chips to China was, in the same season, moving to ban Britain, and every other country on Earth, from using America's best model. Whether that was targeted retaliation against one company or a genuine national security reflex applied without much discrimination, Ball wrote, the result looked the same either way.

Gary Marcus, a longtime industry critic from a different corner of the debate entirely, warned that the directive would give Chinese-born researchers at American labs a fresh reason to go home, and would teach investors that an American AI company's product could be switched off by decisions that had nothing to do with its underlying merit. A number of tech executives and investors raised a narrower but sharper version of the same concern. The three-week gap had handed Chinese open-source developers the same ecosystem examined at length in Chapter 11, valuable time to keep closing a gap that was already closing on its own. An export control action meant to protect American advantage in AI had, in the short run, done more to help the rival ecosystem it was ostensibly guarding against than to slow it down.

The political reaction resisted easy sorting too. Senator Josh Hawley delivered what one account called a blistering rebuke of unfettered AI development around the same period, a Republican pushing for more restriction rather than less. Florida Governor Ron DeSantis separately argued against federal preemption of state AI rules as bad policy. A bipartisan discussion draft, the Great American AI Act, began circulating in the House the same month. None of this cohered into a single clear position on whether the Fable and Mythos suspension itself was justified. What it demonstrated was that the underlying question, how much control the government should exercise over which AI capabilities the country's companies are allowed to ship, had stopped sorting cleanly along the lines that usually organize tech policy fights. Shortly after, OpenAI's equity proposal made the same point from an entirely different angle.

One consultant, describing the shift for corporate technology buyers, reached for language used in an entirely different context. This is not just about data sovereignty anymore, he said. It is about capability sovereignty, where governments want to control who has access to frontier AI capabilities irrespective of who built it, where it is hosted, or who worked on it. That is Chapter 12's entire argument restated from the other side of the ledger. Chapter 12 examined nations building their own compute and their own models because they didn't want their AI ambitions to depend on a decision made in someone else's capital. The Fable and Mythos episode is the proof of concept for exactly the fear that motivated them. It took the US government a single afternoon to demonstrate that access to a frontier model anywhere in the world sits downstream of a decision made in Washington.

The reaction outside the US confirmed it. One analyst covering the enterprise technology fallout warned that a pre-IPO government stake in an American frontier lab, layered on top of an episode where Washington had just switched off a product for the entire world with a single letter, would likely prompt other governments to demand analogous arrangements of their own. That in turn would push European and Asia-Pacific buyers to reassess the assumption that an American AI provider was ever a neutral vendor to begin with. For a British or European company whose government built a sovereign compute fund specifically to avoid depending on American infrastructure, the Fable and Mythos episode was not an abstraction. It was the exact scenario that fund existed to hedge against playing out within weeks of the fund's launch.

The Offer

Three weeks after Fable and Mythos came back online, a different kind of leverage arrived from the opposite direction. On July 2 the *Financial Times* reported that OpenAI had proposed giving the US government a 5 percent equity stake in the company, worth roughly $42.6 billion against the $852 billion valuation OpenAI set in a funding round that March. The vehicle Sam Altman proposed was modeled explicitly on the Alaska Permanent Fund, the sovereign wealth structure that invests Alaska's oil royalties and pays residents an annual dividend regardless of whether they ever worked in the oil industry. Altman's argument, made directly to Trump, Commerce Secretary Howard Lutnick, and Treasury Secretary Scott Bessent, was that giving ordinary Americans a financial stake in AI's upside was the fairest way to share gains from a technology that would otherwise concentrate almost entirely in a handful of already-wealthy hands. The proposal envisioned other major American labs, Anthropic, Google, and Meta among them, contributing comparable stakes to the same fund. Whether any of them would agree remained, as of this writing, entirely unresolved.

This wasn't a new idea dressed up for a slow news week. OpenAI had floated a version of it in an April 2026 policy paper describing what it called a Public Wealth Fund, intended to give every citizen, including people with no money in the stock market at all, an automatic stake in AI-driven growth. Trump's February 2026 executive order had already called for the creation of a federal sovereign wealth fund in general terms. The specific instrument, a government equity stake acquired outside the ordinary channels of taxation or regulation, had precedent that predated OpenAI's proposal by nearly a year. In August 2025 the Treasury converted existing CHIPS Act grants into a roughly 10 percent stake in Intel, an $8.9 billion position that made the federal government Intel's largest shareholder, the episode examined previously as the access-guarantee mechanism arriving at the silicon layer.

The same administration had separately negotiated a 15 percent cut, later raised to 25 percent, on some shipments of Nvidia and AMD's China chip revenue as a condition of export licenses, and had taken equity or equity-like positions in MP Materials and other companies the administration considered strategically significant. OpenAI's offer wasn't inventing a mechanism. It was applying a mechanism the government had already normalized at the silicon layer to the model layer directly, this time volunteered rather than imposed.

The timing carried its own significance. OpenAI had confidentially filed paperwork with the Securities and Exchange Commission the previous month, moving toward a public listing that advisers were weighing as early as later in 2026 at a valuation some reports suggested could clear a trillion dollars. A government equity stake negotiated before that listing would fix Washington's position at a valuation OpenAI itself was actively trying to grow past. The same week the equity story broke, SoftBank closed the second $10 billion tranche of a planned $30 billion investment in OpenAI, priced against the same $852 billion figure the government's proposed 5 percent was calculated against. Private capital and public equity were being negotiated over the identical number in the same news cycle, one flowing in through the ordinary channels of venture financing and the other through a mechanism that didn't exist as a live option for any AI company eighteen months earlier.

That distinction, volunteered rather than imposed, is what makes this half of the chapter different from the export control half, and it is worth holding the two up against each other rather than treating them as the same story twice. Anthropic did not choose to have Fable and Mythos suspended. OpenAI chose, on its own initiative, to offer Washington a slice of itself before anyone forced the question. Reporting on the proposal was directly about the calculation behind it. The donation was explicitly meant to secure good relations with the administration and address political blowback in a climate where the same administration had just demonstrated, against Anthropic, exactly how fast it could turn off a competitor's flagship product with a single letter. It was insurance, purchased in equity, against a switch OpenAI had just watched Washington flip on someone else.

The politics of this scrambled the usual lines. Independent Senator Bernie Sanders, who had met with Altman the previous month, introduced his own version days before OpenAI's proposal became public. The American AI Sovereign Wealth Fund Act would impose a one-time 50 percent tax on the stock of any AI company generating more than $200 million a year in AI-related revenue, funneling the proceeds into a public fund modeled on the same Alaska precedent. Sanders's framing was pointed: AI systems are built on the collective output of human knowledge, books, scientific research, journalism, and millions of ordinary conversations, extracted without permission or compensation from the people who created them. The public deserves a direct ownership stake in what that extraction produced, not just a tax bill paid after the fact.

Steve Bannon, the former Trump chief strategist, reached almost the identical conclusion from a completely different set of premises, publicly dismissing OpenAI's 5 percent offer as tip money and demanding the government take 50 percent of AI companies' equity outright. A socialist senator from Vermont and a nationalist former White House strategist arrived, within the same news cycle, at the same number, for reasons that shared almost nothing else in common. When the far left and the populist right independently converge on identical terms, that's rarely a coincidence of two isolated arguments. It's usually a sign that a number everyone else is treating as generous, 5 percent, looks to enough of the political spectrum like the opening bid in a negotiation rather than the final offer.

Anthropic's response to the same pressure was to decline the specific mechanism OpenAI embraced. According to reporting in the days following the *Financial Times* story, Anthropic was not participating in equity discussions with the administration at all. The company had, in its own policy paper released the previous month, proposed a different answer to the same underlying problem. Universal predistributive capital accounts, giving every American a direct financial stake in economic growth at the moment AI begins driving it, with priority given to workers displaced by the technology first. Two labs, facing the identical political pressure to show the public it would share in AI's gains, reached for two different instruments. OpenAI offered equity in itself, directly, to the government. Anthropic proposed a broader account structure that didn't require handing Washington a board seat. Which of those turns out to matter more may depend less on economic logic than on which lab is negotiating from a position of having just watched a competitor's product get switched off.

What the Switch Reveals

Watchdog groups raised the objection this chapter would be negligent to skip past, and at least one put it in terms precise enough to quote directly. Nat Purser, a senior policy advocate at the Washington nonprofit Public Knowledge, framed the structural problem plainly. The government would be a shareholder and a regulator at the same time, which creates substantial conflicts of interest. The public should not want a situation where the government becomes less willing to impose or enforce safety rules because doing so could reduce the value of its own investment. The concern isn't hypothetical or distant. If Washington holds tens of billions of dollars in OpenAI equity, every future safety regulation, antitrust action, or data protection rule that dents OpenAI's valuation also dents a line item on the government's balance sheet. That is the textbook definition of regulatory capture, except arriving through a voluntary donation rather than the usual channels of lobbying and revolving doors.

The legal basis for any of this remains considerably murkier than the confident headlines suggest. Legal scholars examining the government's existing equity positions, Intel among them, found no statute that expressly authorizes Washington to take stakes in private companies as a matter of course. The Trump administration has never published a detailed legal justification for the ones it already holds. The most plausible authority is broad contracting language stretched to cover something it wasn't obviously written for, alongside processes like the national security review that produced the government's stake in U.S. Steel. The practical reason none of this has been tested in court is simpler than the legal theory. So long as a company formally consents to the arrangement, as Intel did and as OpenAI would be volunteering to do, there's no injured party left standing to sue. Consent doesn't just make the politics easier. It may be the only thing making the whole mechanism legally durable at all.

The politics scrambled further than a simple left-versus-right story once critics started examining Sanders's bill rather than just OpenAI's offer. Bruce

Schneier, a security technologist with no particular allegiance to the administration, argued that Sanders's approach would leave control in exactly the same hands it started in. Control would be maintained by the tech oligarchs, he said. The only difference is that the government will now have a conflict of interest when it comes time to regulate them. He proposed taxing AI companies directly and building a separate, public AI option outside the for-profit market entirely rather than taking equity in the incumbents. Tad DeHaven, a Cato Institute analyst with every institutional reason to distrust Sanders's instincts, made the opposite-facing but structurally identical point. Sanders wants political control over AI companies through voting shares and board seats, DeHaven wrote, and Trump's pursuit of government equity stakes in Intel and elsewhere had already opened the door for exactly that ask. A libertarian critic and a security-focused technologist, coming from unrelated premises, converged on the same warning Purser had raised from the center-left. Once government equity in private AI companies becomes a normal tool, the fight over how much equity and with what governance stops being a fight about AI at all. Instead it becomes a fight about how much of the American corporate sector the state ends up holding a stake in.

None of this was happening in a political vacuum. Public opinion data from the same season showed why the pressure existed in the first place. A majority, actually 55 percent of Americans, told Quinnipiac pollsters that they believed AI would do more harm than good in their daily lives. In another poll 71 percent told YouGov they thought AI was moving too fast for anyone, regulators included, to keep up. The pressure wasn't confined to Washington either. California Governor Gavin Newsom signed an executive order in May 2026 directing state agencies to evaluate whether residents should receive direct ownership stakes in AI companies or the funds profiting from them, a state-level echo of the same demand running through Sanders's bill and OpenAI's proposal, arriving before either had been resolved at the federal level.

OpenAI's product decisions were already bending to the same pressure well before the equity story broke. The company restricted the rollout of its GPT-5.6 models to a limited set of clients at the government's specific request, a smaller and quieter version of the Fable and Mythos suspension that barely made headlines precisely because OpenAI complied before Commerce needed to send a letter at all. Whatever mechanism eventually prevails, a 5 percent donation, a 50 percent tax, or something Congress hasn't drafted yet, the underlying political demand it's responding to isn't going away on its own.

Both halves of this chapter are versions of the same mechanism applied at every previous layer before finally reaching the model itself. The Indispensability Trap predicts that something essential enough eventually gets its returns capped or its access conditioned by the society that depends on it. Fable and Mythos show the imposed version of that mechanism arriving in an afternoon, using legal authority stretched so far past its original design that a federal court held it to be illegal. OpenAI's offer shows the voluntary version arriving first, a company choosing to hand over a piece of itself rather than wait to discover what an unfavorable version of the same leverage might cost it later. Jevons paradox is present in the background of both. The cheaper and more capable frontier models become, the more of them get built into critical infrastructure. The more of them getting built into critical infrastructure is exactly what gives an export control letter or an equity demand its teeth. The Perez framework's installation frenzy explains why OpenAI moved first and alone. In a build-out this capital-intensive, the surest way to protect a lead is to make peace with the referee before a competitor does it first, or before the referee stops asking nicely.

None of this is settled, and while there is no neat bow to end this chapter, this is the state of things. The export control authority Commerce used exists on paper but was never written for anything close to a blanket worldwide ban. And now a federal court has said it was unlawful. Unlawful or not, it certainly put a chill on Anthropic's use in the federal market. The equity proposal remains, as of this writing, a conversation rather than a signed agreement, and would likely need an act of Congress to formalize. Whether other labs follow OpenAI's lead, whether Sanders's bill or something like it passes, whether Commerce writes a rule that makes the next Fable-and-Mythos episode predictable instead of improvised, are all open questions I cannot answer. The people fighting them out haven't answered them yet either. What's no longer in question is that the switch exists, in both directions, and every lab now knows it.

The eight companies mapped back in Chapter 7 have spent six chapters making increasingly desperate, increasingly intertwined bets about which layer of this stack will hold durable value. The model itself has been commoditized out from under them. This chapter adds a variable none of those bets fully priced in. The state itself deciding, sometimes with a letter and sometimes with a stock certificate, who gets to keep playing. Where those seven bets actually land, now that every force traced is on the table at once, is the question Part III's final chapter will answer.

Chapter 14

Thirty Days from Going Out of Business

Our company is thirty days from going out of business.

—Nvidia's unofficial company motto, 1997

In the spring of 1993, three engineers sat in a booth at a Denny's in San Jose and talked themselves into starting a chip company. Jensen Huang had been running a division at LSI Logic. Chris Malachowsky and Curtis Priem were at Sun Microsystems. The idea was that the personal computer was about to become a machine for playing three-dimensional games, that rendering those games would require a kind of math general-purpose processors did badly, and that a company which built a chip to do that math well could sell it to every gamer on Earth. They named the company after the Latin word for envy. They raised their first real money from Sequoia two years later. And then they very nearly died.

The first product, a chip called the NV1, shipped in 1995 and was a technical marvel that solved the wrong problem. Nvidia had bet on rendering curved surfaces from four-sided shapes at the exact moment the rest of the industry, and Microsoft's Direct3D standard, settled on triangles. Nobody wanted it. A partnership with Sega to build the graphics chip for the Dreamcast console consumed a year on the same doomed approach before Sega pulled the plug and switched vendors. Sega, in an act of generosity that ranks among the more consequential in the history of the industry, invested $5 million in the company it had just fired, which was enough to keep the lights on. Nvidia laid off half of its hundred employees and bet everything on one more chip.

That chip was the RIVA 128, and the way Nvidia built it tells you most of what you need to know about the company. Huang abandoned the proprietary approach entirely, adopted the industry standard everyone else was already using, and set out to execute it faster than anyone. By the time the RIVA 128 shipped in August of 1997, the company had one month of payroll left. It sold a million units in four months. Somewhere in that stretch, the phrase that became the company's unofficial motto was born, and Huang has repeated it for nearly thirty years, long after it stopped being literally true: Our company is thirty days from going out of business.

Nvidia went public in January of 1999 at a market value of about $563 million. That fall it shipped the GeForce 256 and, in a piece of marketing that turned out to be a piece of history, called it the world's first GPU. On the day I am writing this, Nvidia is worth about $5.4 trillion, which is more than any company has ever been worth, and its stock has risen roughly one million percent since that IPO. It is the most indispensable company on Earth by any measure this book has used, and for exactly that reason it is the company the book's argument has the hardest time with. Every prior chapter says the company in Nvidia's position does not keep the fortune. Nvidia is keeping it, spectacularly, and doing things with it that no infrastructure builder in any of the four earlier build-outs ever managed. This chapter is about whether that is the exception that breaks the pattern or the most elaborate version of the pattern yet.

The Bet Nobody Wanted

The company that became the beating heart of the artificial intelligence economy did not get there by planning to. It got there by making a bet in 2006 that almost nobody on Wall Street understood and that took six years to pay off.

A graphics chip renders a scene by doing the same small calculation millions of times at once, one for every pixel. That is a different shape of problem from the one a conventional processor is built for, which is doing one complicated thing after another as fast as possible. By the early 2000s, a handful of researchers had noticed that the parallel shape of graphics math was also the shape of a great deal of scientific computing, and that a gaming card was, if you could figure out how to program it, a cheap supercomputer. Programming it was the hard part. You had to disguise your physics problem as a graphics problem and trick the chip into solving it.

In 2006, Nvidia released CUDA, a software platform that let developers write ordinary code and run it on the GPU's thousands of parallel cores without the disguise. It was, at the time, a strange thing for a gaming-chip company to do. CUDA was a cost center. It consumed engineering talent and shipped on every consumer graphics card whether or not the buyer ever used it, which nearly all of them did not. For years, analysts asked why a company that made its money from teenagers playing Crysis was spending so much on a compiler for physicists. The company built the software ecosystem before the demand existed, which is the oldest and most dangerous move in this book, the animating faith of the fiber boom, and the reason it worked this time is the reason this chapter exists.

The demand arrived on September 30, 2012, and it arrived from a direction nobody at Nvidia had planned for. Three researchers at the University of Toronto, Alex Krizhevsky, Ilya Sutskever, and Geoffrey Hinton, entered a deep neural network called AlexNet into an image-recognition competition called ImageNet and crushed the field by a margin that ended the argument about whether neural networks were a dead end. They had trained it on two consumer GTX 580 gaming cards, bought off the shelf, running CUDA. What had taken months on a rack of conventional processors took days on two cards a teenager could buy at Best Buy. Every serious researcher in the field read the paper, understood what the two cards meant, and went out and bought Nvidia hardware. The company had spent six years building a road to nowhere in particular, and the most important technology of the century turned out to be waiting at the end of it.

Huang understood what had happened faster than anyone else in the industry, and the next decade of Nvidia's history is the record of a company reorganizing itself around a customer that barely existed yet. In August of 2016, Huang personally delivered the first DGX-1, a purpose-built deep-learning machine the company described as an AI supercomputer in a box, to a small nonprofit research lab in San Francisco called OpenAI. He signed it. A photograph of the delivery survives, and looking at it now is like looking at a photograph of Samuel Insull handing a light bulb to a farmer. The company that made the grid was seeding the appliance makers before there was a grid.

Two more decisions in that decade matter for what follows. In 2019, Nvidia agreed to buy Mellanox, an Israeli networking company, for $6.9 billion. Mellanox made the interconnects that let thousands of chips talk to each other fast enough to act as one machine, and without that plumbing the largest AI clusters in the world could not exist. It may be the most important acquisition in the history of the AI build-out, and almost nobody outside the industry noticed it. The other decision was the one that didn't happen. In 2020 Nvidia agreed to buy Arm for $40 billion, the British chip designer whose architecture sits inside nearly every phone on Earth. Regulators in Washington, London, Brussels, and Beijing all objected, and in February of 2022 Nvidia walked away. Hold that one. It is the first time in this story the sword swung at Nvidia, and it will not be the last.

Scarcity

On November 30, 2022, OpenAI released ChatGPT, and the customer Nvidia had spent a decade cultivating became, almost overnight, the largest buyer of anything in the history of the technology industry.

The numbers that followed do not read like the financial history of a company. They read like the financial history of a country. Nvidia's market value crossed $1 trillion in May of 2023, $2 trillion in February of 2024, $3 trillion that June, $4 trillion in July of 2025, and $5 trillion in October of 2025, the first company ever to reach that number. Its revenue for the fiscal year that ended in January of 2026 was $215.9 billion, of which $193.7 billion came from data centers, a business that had been a rounding error a decade earlier. In the quarter it reported at the end of August 2026, as I was finishing this book, revenue was $96.2 billion, up 106 percent from the year before, and data center revenue alone was $89 billion. The gross margin on all of that was 75 percent. The company told investors to expect $108 billion the following quarter and said it had visibility to $1 trillion in orders across its current and next generation of chips. Its chief financial officer, Colette Kress, said the five largest cloud companies were on course to spend $800 billion on capital projects in 2026 and $1.3 trillion in 2027, a substantial share of it on Nvidia's products. Amazon alone committed to buying two million Nvidia GPUs.

The mechanism underneath those numbers is the one this book has been describing since the railroad chapter, and it is worth stating in Nvidia's terms.

The product was scarce. Not scarce in the way a luxury good is scarce, by design, but scarce the way a railroad line into a boomtown is scarce, because there is only one and everyone needs it at once. Chapter 9 traced the physical reason, the single Taiwanese company and the single packaging process that every advanced chip on Earth has to pass through. Nvidia sat on the demand side of that chokepoint holding the only product the buyers wanted and the only software they knew how to use, and it priced accordingly. A 75 percent gross margin is what indispensability looks like on an income statement.

What makes the scarcity durable, so far, is CUDA. Every AI engineer on Earth learned to write for it, every framework was built on it, and switching a large model to another company's chips means rewriting years of accumulated work. This is a moat made of other people's labor, the same kind of moat the Bell System had when every telephone in America plugged into its network, and it has held longer than any competitor expected. But it is eroding, and the erosion is measurable. Nvidia's share of the market for AI accelerators was roughly 92 percent in 2023. By 2026 it was somewhere between 80 and 85 percent. Google's own chip, the TPU, is in its eighth generation and shipping by the millions. AMD has built a real business at 5 to 8 percent of the market. Every hyperscaler in this book is designing its own silicon for the specific purpose of needing fewer of Nvidia's. That is what happens to every grid in this book. The customers who depend on you most are the customers with the strongest incentive to stop.

Huang has known this since the days when the company's survival was measured in weeks, and his response to it is the reason Nvidia is the most interesting company in the book.

The Balance Sheet as a Weapon

Somewhere in 2024, Nvidia stopped behaving like a supplier and started behaving like the central bank of the artificial intelligence economy.

The evidence is in the company's own filings. In the quarterly report it filed at the end of August 2026, Nvidia disclosed $42.8 billion in marketable equity securities and $51.2 billion in non-marketable ones, which is to say stakes in private companies, with roughly $18 billion in additional commitments still to fund. In the fiscal year that ended in January, it had put $17.5 billion into private companies and infrastructure funds. It did sixty-seven venture deals in 2025 alone. The inventory of what it owns reads like a map of the industry this book has spent seven chapters describing.

It owns a multi-billion-dollar stake in OpenAI, the largest of its bets, scaled down from a framework announced in the fall of 2025 that would have reached $100 billion had it ever been built as described. It committed up to $10 billion to Anthropic in February of 2026, in the same round where Microsoft committed $5 billion. It put $10 billion into xAI, which became roughly $21 billion of SpaceX stock when SpaceX absorbed xAI, making Nvidia the sixth-largest outside investor in a rocket company that builds its data centers exclusively on Nvidia chips. It owns 7 percent of CoreWeave, the neocloud Chapter 8 examined, alongside a $6.3 billion commitment to buy back CoreWeave's unsold capacity. It put $2 billion into Nebius, another neocloud. It bought 4 percent of Intel for $5 billion, the rescue Chapter 9 already examined from the other side. It put $5 billion into Safe Superintelligence, $1 billion into Naver, roughly $2 billion each into Marvell, Synopsys, Coherent, and Lumentum, and it has been a repeat investor in Mistral, Cohere, and Perplexity.

Huang's explanation for all of this, offered on a podcast in April of 2026, is the most candid description of the strategy anyone has given. There are so many great foundation model companies, he said, and we try to invest in all of them. We don't pick winners. We need to support everyone.

Read that sentence the way this book reads everything, as a statement about where the money goes. A supplier that invests in all of its customers is not making venture bets. It is making sure that every company capable of buying its product has the capital to keep buying it, and that the money it collects from them comes back around as equity in the buyers. Chapter 8 called this circular financing when the amounts were smaller and the participants fewer. Nvidia has industrialized it. It invests in the labs, the labs buy Nvidia chips through clouds that Nvidia also invests in, and the clouds' unsold capacity is guaranteed by Nvidia's own promise to buy it back. Every dollar in that loop passes through Nvidia's income statement on the way around, at a 75 percent margin.

Then there are the acquisitions, which are the balance sheet used not to support the ecosystem but to absorb it. On Christmas Eve of 2025, Nvidia paid roughly $20 billion in cash for the assets of Groq, a startup founded by the engineers who had designed Google's TPU, whose inference chips were one of the few credible threats to Nvidia's position in the fastest-growing part of the market. It was the largest transaction in the company's history, nearly three times what Groq had been valued at three months earlier, and it was structured as something other than an acquisition. Groq called it a non-exclusive licensing agreement. Groq's founder, its president, and its core engineering team joined Nvidia, and a legal entity called Groq continued to exist independently, running a cloud service. The deal was announced and closed within the same week. There was no merger review, because on paper there was no merger. A senator from Massachusetts and a senator from Connecticut sent a letter in March asking how that could possibly be true, and the Federal Trade Commission has begun looking at the structure, which the industry has taken to calling a reverse acquihire. Nvidia had learned something from the Arm episode. If regulators will not let you buy the company, buy everything about the company except the part that triggers the review.

And then, on the second of September 2026, days before I finished this manuscript, Nvidia agreed to pay $12.9 billion for Hugging Face.

If you have never heard of Hugging Face, you have never trained a model, and if you have trained a model, you have almost certainly used it. It is the platform where the open-weight world lives, eighteen million developers, three million models, half a million datasets, two hundred thousand companies, the shared library and town square of everything Chapter 10 and Chapter 11 described. The Chinese models that crossed over American ones in download share were counted on Hugging Face. Mistral's weights live there. So do Nvidia's own, more than five hundred of them, which had already made the chipmaker the single largest contributor of open models to the platform before it decided to buy the platform outright. Hugging Face had turned down a $500 million investment from Nvidia the year before, at a $7 billion valuation, because its founders wanted to stay independent. What changed was a security breach over the summer that convinced Clem Delangue, the company's chief executive, that open-source AI had reached a turning point and needed more resources than an independent company could provide. He approached Huang. Huang, in the blog post announcing the deal, promised the platform would remain open to the entire ecosystem and that Nvidia's compute would not be required to use it.

I believe him, and I also think the promise misses the point. Nvidia did not buy Hugging Face to force anyone onto its chips. It bought Hugging Face because a thriving open-model ecosystem is the best insurance policy Nvidia can own against the closed labs. Every one of the large, closed labs—OpenAI, Google, Anthropic, Amazon—is building its own silicon for the specific purpose of buying fewer of Nvidia's. The open-weight world, by contrast, runs on whatever hardware is cheapest and most available, which for now means Nvidia. A company that makes it easier for the rest of the world to build on open models is a company that keeps more of the world's AI running on hardware the closed labs cannot substitute. Chapter 10 argued that the base model was commoditizing. Nvidia just bought the commodity exchange.

Compute as an Asset Class

Everything so far, the stakes, the buybacks of unsold capacity, the acquisitions, is a company using its balance sheet to shape the market it sells into. What Nvidia did in the summer of 2026 is different in kind, and it is the move that ties this company to the deepest pattern in this book.

In July 2026 Nvidia launched a program it calls DSX AI Factories, which the next chapter will examine as one bet among many. The mechanics are worth restating here because of what came next. A cloud provider or a startup can draw on Nvidia's chips without paying the full price upfront. In exchange Nvidia takes a recurring share of whatever revenue that capacity later earns, and it backstops the whole arrangement with a guarantee to repurchase any capacity that goes unsold. Nvidia's own explanation for why the program exists was unusually frank. Even long-term customer commitments, the company said, had proved insufficient to unlock financing for large-scale computing deployments. The most dominant supplier in the industry was admitting that the credit markets no longer believed the demand story on their own, and that Nvidia would have to carry the belief on its own balance sheet.

Then, on August 10, Nvidia announced that it would not carry it alone. It signed memoranda of understanding with Apollo, BlackRock, Blackstone, Brookfield, Goldman Sachs, and KKR, six of the largest pools of capital on Earth, to establish what the press release called independent compute financing platforms intended to mobilize more than $500 billion of third-party money for the build-out of AI infrastructure over time. The headline on the release was the honest part. The new platforms, it said, would turn Nvidia compute into an investable asset class.

Huang's own words in that release deserve to be quoted, because they are the clearest statement any infrastructure builder in this book has ever made about what he is doing. We began by building chips, he said. Today, we are helping create a new class of productive, investable infrastructure: AI factories. And then the sentence that should be engraved over the entrance to this chapter: In AI, compute is revenue.

Consider what that means. A GPU is a piece of production capital. It sits in a rack and does work. What Nvidia announced in August was a machine for converting that production capital into financial capital, into a security that a pension fund can buy, that pays a yield linked to how much the chip gets used, that can be pooled and tranched and sold on to someone else. The company that makes the most important physical asset of the fifth build-out is now in the business of manufacturing the paper that finances it. And the paper has Nvidia's own guarantee behind it. In the same quarterly report that disclosed the equity stakes, Nvidia disclosed that it had entered into guarantees capped at $105 billion to provide credit support for a data center project in Ohio being built for an affiliate of OpenAI, about 4.25 gigawatts of computing load, nearly the size of Meta's Hyperion project described in Chapter 1. One company's balance sheet, standing behind one customer's data center, for a sum larger than the annual revenue of all but a handful of corporations on Earth.

There is a name for a company that does this, and Chapter 2 introduced it without needing this part of its history. The name is General Electric.

GE's climb from dynamos to light bulbs and motors and appliances is the one case in this book where an infrastructure builder kept the fortune, and every full-stack aspirant in Part III is trying to repeat it. But GE's story has a second half that Chapter 2 did not need and this chapter does. Somewhere in the middle of the twentieth century, General Electric noticed that its customers often could not afford what it sold them, and it began financing the purchases itself, first appliances, then locomotives and jet engines, then everything. The financing arm was called GE Capital, and for a long time it was the best business the company had ever built. It let GE sell more of everything. It earned a spread on every sale. It turned the company's balance sheet into a weapon its industrial competitors could not match. By 2007 GE Capital produced roughly half of General Electric's profits, and the most admired industrial company in America was, underneath its logo, a bank.

In 2008 the bank nearly killed the company. When the credit markets froze, GE Capital could not roll its short-term debt, and General Electric, the company that had invented the modern industrial corporation, needed a $3 billion rescue from Warren Buffett and a federal guarantee on its debt to stay liquid. It cut its dividend for the first time since the Depression. It spent the next decade selling GE Capital off in pieces, and when that was done, the conglomerate itself was broken into three companies. The balance sheet as weapon is the most powerful move an infrastructure company can make. History's clearest example of it ended with the weapon pointed backward.

I am not predicting that ending for Nvidia, and the differences matter. General Electric financed its customers with borrowed money. Nvidia is financing its customers out of a cash flow that Bank of America estimates at more than $400 billion across 2026 and 2027, and its total equity exposure, roughly $70 billion, is a fraction of that. A company that can lose every stake it holds and still return half its free cash flow to shareholders, as Nvidia has committed to doing, is not in GE Capital's position. But the $500 billion platforms are not Nvidia's own money. They are other people's money, raised on the premise that a GPU's usage-linked revenue is a stable, long-duration asset, and the premise is being tested in the market right now. CoreWeave, the neocloud Nvidia owns 7 percent of and whose unsold capacity it has promised to buy, saw its credit-default swaps price at roughly even odds of default in the summer of 2026, with an interest bill larger than its entire 2025 revenue. That is the first crack in the asset class, and it appeared before the asset class formally existed.

The Sword Has Already Swung Three Times

Every chapter in Part II ended the same way, with the state arriving to cap the returns of the company that had become essential. The instinct is to ask when that will happen to Nvidia. The more accurate question is why it has not been noticed that it already has.

The first swing was Arm, in 2022, and it is the clean case. Nvidia tried to buy the most important chip architecture on Earth, and four governments said no, and it walked away. That is the antitrust sword exactly as Chapter 6 traced it from IBM to Microsoft to Google, applied preemptively to a merger rather than after the fact to a monopoly.

The second swing is stranger, and it has no precedent in the four earlier buildouts. Beginning in 2022, the United States government told Nvidia which of its products it could sell to its largest foreign market, and the answer kept changing. The A100 and H100 were banned for China in the fall of 2022, so Nvidia designed a compliant chip. That chip was restricted, so it designed another. The H20, the third such chip, was restricted in April of 2025, costing Nvidia a write-down in the billions, then unrestricted that July after Huang lobbied the administration personally, then superseded in December by permission to sell the more powerful H200, on one condition that no exporter in American history had ever been asked to meet. The United States Treasury would take 25 percent of the revenue on every chip sold. Not a tariff on the buyer. A share of the seller's sales, paid to the government, as the price of a license.

Chapter 4 watched the Justice Department take the Bell System's future in 1956. Chapter 5 watched rate commissions cap the returns of the utilities. None of them ever simply took a percentage of the top line. The revenue share is the Indispensability Trap's sword refashioned as a tollbooth, and it was accepted by the most valuable company on Earth because the alternative was a market Huang estimated at $50 billion a year vanishing entirely. It then vanished anyway. Beijing, having watched Washington set the terms, blocked the imports on its own side to push Chinese buyers toward Huawei's domestic chips, which were unveiled in March of 2026 and entered mass production the following month. By July a Commerce Department official was telling Congress that shipments under the $10 billion in approved licenses remained minimal. Nvidia's China revenue, which had been a fifth of the company's total, is now less than 1 percent of its data center business. Two governments, acting in opposition, removed the world's second-largest AI market from the world's most indispensable AI company, and the company kept growing at triple digits anyway. That is how much scarcity elsewhere was worth.

The third swing is the one still in the air. The Groq structure, an acquisition designed to look like a license, drew the Senate letter and the FTC's attention within three months. The Hugging Face deal is expressly conditioned on regulatory approval that will not come before 2027, and it hands the most dominant hardware company in the industry the most important software platform in the open-model ecosystem. A House committee has advanced legislation that would treat AI chip exports the way the law treats arms sales, with congressional review of each. The pattern from every earlier build-out is that the sword arrives after the abuse, once the indispensability is established and the public has noticed. Nvidia's indispensability was established years ago. The public has noticed. What is different this time, and it is the same difference found in the sovereign AI hedges and in OpenAI's equity offer discussed in earlier chapters, is that the state is not waiting for the abuse. It is taking its share in real time, chip by chip, deal by deal, in a negotiation that never formally ends.

The Climb

Set the state aside and look at what Nvidia has built, because the shape of it is the answer to the question this chapter opened with.

At the bottom sit the chips: the GPUs, the Grace and Vera processors that now ship alongside them, and, since Christmas, Groq's inference silicon. One layer up sits the networking, Mellanox's inheritance, the InfiniBand and NVLink fabric that makes a hundred thousand chips act as one machine. Above that sit the systems, the DGX boxes and the rack-scale Blackwell and Vera Rubin platforms that arrive as a finished computer rather than a component. Above the systems sits the software, CUDA and the libraries and the inference microservices Nvidia calls NIM, the layer every developer touches and the reason the chips are so hard to replace. Above the software sit Nvidia's own models, the Nemotron family, hundreds of open models the company gives away for exactly the reason Meta and Mistral do. Above the models, as of this month, sits Hugging Face, the distribution platform where the models live and where the developers go to find them. And beside the whole stack, financing every layer of it, sit the DSX program and the $500 billion platforms. Beyond the data center, the same architecture is being pointed at robots and factories, a physical-AI business that Nvidia has now disclosed contributing more than $6 billion in fiscal year 2026. An analyst described it as a five-layer cake. It is more layers than that, and the frosting is money.

This is the General Electric climb, from the dynamo to everything the dynamo powers, and Nvidia is further up it than any company in Part III. Google owns models and chips and distribution but rents the interconnect story from Nvidia. Microsoft owns distribution and models but builds on Nvidia's silicon.

Oracle owns nothing but the buildings. Nvidia owns the chip, the network, the system, the software, an open-model franchise, the platform the open models live on, and now the financing. It has climbed from the grid into the appliance business, into the appliance store, and into the bank that lends the customers the money to shop there.

Whether the climb holds depends on the same question the book has asked of every climber, and Chapter 10 gave the uncomfortable answer for everyone else: Every layer commoditizes eventually, and the value keeps moving to the one above. Nvidia's answer is not to pick a layer. It is to own enough of them that when value migrates it migrates into another room of the same house. A supplier whose customers are building their own chips holds the customers with its software. A software franchise that competitors are trying to replicate is defended by owning the models developers actually run. A model business that is commoditizing by design is monetized by owning the platform where commoditized models are traded. And every one of those layers is financed by a program that pays Nvidia a share of the revenue whether or not the layer itself proves defensible. It is the most complete answer to the Indispensability Trap that any company in five build-outs has constructed, and the fact that it needed to be this complete is the tell.

Four Engines, One Company

Nvidia is the only company in this book on which all four of its analytical engines run at once without any of them needing to be adjusted.

The Indispensability Trap is present in its purest form: A company that became essential, priced its essentialness at a 75 percent margin, and has already had a percentage of its sales claimed by the state, its largest acquisition blocked, its second-largest structured to avoid review, and its largest foreign market removed. Every mechanism the earlier chapters found has touched Nvidia within four years, and the company is still growing at rates no business its size has ever sustained. The trap is real and it is not, so far, decisive. The question the book asks of every build-out is who keeps the money, and the honest answer in Nvidia's case is that, so far, Nvidia does.

Jevons paradox explains why. CUDA made a kind of computing cheap that had been prohibitively expensive, and cheap parallel computing did not shrink the demand for it. It created industries. Every subsequent cost collapse Nvidia engineered, and Chapter 10 documented the inference-cost collapse in detail, increased the total amount of money spent on Nvidia's products. Compute is revenue, in Huang's phrase, precisely because each time compute gets cheaper the world finds more revenue to run on it. Nvidia is the company that has understood Jevons better than any other in this book, and its entire financing apparatus is a bet that the paradox keeps holding.

The Perez framework is where the story gets uncomfortable. Every installation phase in this book was financed by capital that ran ahead of confirmed demand, and every one of them had a vendor who financed its own customers' purchases when the ordinary lenders got nervous. Chapter 6 found Lucent and Nortel doing it in the fiber boom, extending credit to carriers who used it to buy Lucent and Nortel equipment, and it found the bill arriving when the carriers went bankrupt and the vendors were left holding paper that a demand curve had promised to redeem. Nvidia's DSX program and its $500 billion platforms are that vendor financing at a hundred times the scale, with the difference that Nvidia has the cash to survive its customers' failures and Lucent did not. That difference is real. It is also exactly the difference General Electric believed it had.

The distinction between financial and production capital cuts straight through the middle of this company, and it is the question the next chapter will have to weigh. The chips are production capital in the purest sense, machines that do work whether or not anyone's stock price holds. The $94 billion book of securities, the $105 billion of guarantees, the repurchase promises, the platforms that turn a GPU into a yield, are financial capital, and Nvidia is manufacturing more of it every quarter. In every earlier build-out, the two kinds of capital lived in different companies, the builders on one side and the financiers on the other, and the crash sorted them. Nvidia has put both inside one balance sheet, on purpose, as a strategy. If the demand curve holds, it is the most brilliant piece of corporate construction since Insull's holding companies. If the demand curve does not hold, it is the same construction, and the book already shows how that one ended.

A company that has spent thirty years telling itself it is thirty days from going out of business has built the one machine in this story designed to make sure that it never is. Whether the machine works, whether the fortune stays where

Nvidia has put it or migrates one last time, up and out of the grid to whoever treats even Nvidia's compute as a free input, is the question the final chapter of Part III now has to answer for every company on the map at once.

Chapter 15

Where the Bets Land

> In the long run, we are all dead.
>
> —John Maynard Keynes, *A Tract on Monetary Reform*, 1923

In a sixty-day window across the spring and early summer of 2026, five companies that compete with each other on almost everything else arrived, independently, at the identical conclusion. OpenAI and Anthropic launched forward-deployed engineering units in May. Amazon committed a billion dollars to the same practice on June 30. Microsoft answered two days later with Frontier Company, $2.5 billion and six thousand people, embedding engineers directly inside client operations rather than simply selling them a subscription. None of them were copying each other in any formal sense.

All of them were copying Palantir, the data-analytics company that had spent two decades building its reputation on exactly this model. Palantir embedded its own engineers inside intelligence agencies and, more recently, partnered with Nvidia to bring that same embedded approach to classified and air-gapped government deployments running Nvidia's open models. It did all of this long before any of the five companies in this chapter treated the practice as a strategy worth naming.

All five had reached, on their own separate timelines, the same reading of the argument made in Chapter 10: The model itself is not the defensible layer anymore. Satya Nadella didn't pull any punches when he wrote in a June essay, reportedly reaching some sixty million views before Frontier Company even had a name, that no single model provider can count on model quality alone as a durable revenue source. The models are becoming interchangeable too quickly. They are commoditizing just the way the premise of the Indispensability Trap says they will. The team standing in the customer's building when the system finally works, not the lab that trained the weights, is what captures the institutional trust, the workflow data, and the switching costs that make the vendor too expensive to lose.

That convergence, five rivals reaching for the identical hedge inside two months without coordinating, is the closing argument I have been assembling since Chapter 7. It is also, uncomfortably for at least one of the five, not yet proof of anything working.

Microsoft's stock fell roughly 21 percent over the first half of 2026, by far the worst performance among the mega-cap technology companies, with June being the company's worst month since December 2000. Frontier Company was announced into that decline, not ahead of it. A hedge this widely and this suddenly adopted is either the market correctly pricing in where value is migrating next or a great deal of capital rushing toward the same unproven exit at once. This book, and this time, cannot tell you which.

What it can do is take the eight-company map and test each of its bets against everything Part III has found and reported. One bet nobody had fully priced in when this book was envisioned has joined the map since, and a new kind of player entirely. We will examine it in its own right at the end of this chapter. But it has started deciding who gets to keep playing at all. This is where they all stand.

The Full-Stack Bet: Google and Microsoft

Google is the only company in this book's entire map that uses the words "full-stack approach" in its own annual filing with the Securities and Exchange Commission, and it is not exaggerating. It owns Google DeepMind, the lab that builds the Gemini models; its own silicon, the Tensor Processing Unit line now in its eighth generation, split into a training-optimized variant and an inference-optimized variant, with roughly 4.3 million units projected to ship in 2026 alone according to semiconductor industry analysts; the cloud infrastructure to run all of it; and, through Search, Android, Chrome, YouTube, and Workspace, a distribution network no competitor can match. The TPU is the only non-Nvidia accelerator shipping at a similar kind of unit volume. AMD's Instinct line has built a smaller but real business of its own, several billion dollars a year at roughly 5 to 8 percent of the market, while Intel's rival Gaudi line has not. The Gemini app passed 900 million monthly users in May 2026. This is the GE bet Chapter 2 described, the wager that a single company smart enough and large enough to own every layer at once escapes the fate of the pure-play infrastructure owner. Nvidia may be the closest thing that the AI era has produced to General Electric climbing from dynamos into light bulbs and motors, but Google is a close second.

Microsoft is making the identical bet from a different starting position, and Chapter 8 already documented how uncomfortably that position must bend to hold. Microsoft owns roughly 27 percent of OpenAI, sells Anthropic's Claude to its own enterprise customers through Copilot Cowork, and ships its own in-house MAI models, three rival sources of frontier intelligence running through a single product line because no single one of the three felt safe to depend on alone. Frontier Company is the same instinct's newest layer, an explicit bet that owning the deployment relationship, the embedded engineers who understand which data feeds which model and which process broke before it started working, is worth more than owning the model running underneath.

Judson Althoff, who runs Microsoft's commercial business, made the pitch to enterprise buyers directly. Microsoft supports more models, more connectors, and more integrations with open systems of record than a company like Palantir, positioning itself as the neutral integrator rather than a single-vendor lock-in. Whether that neutrality is real or a sales pitch is precisely the tension found in the same company's identical hedge some chapters ago. What's changed is the stock market's verdict, at least so far. Microsoft's shareholders spent the first half of 2026 pricing in doubt, not confidence, about whether any of this converts into revenue commensurate with the capital behind it.

The Federal Trade Commission investigation into whether the Microsoft-OpenAI structure was built specifically to avoid the merger review a straightforward acquisition would have triggered, is still open as this book goes to press. Frontier Company adds a new wrinkle to it rather than resolving anything. A company that already holds effective influence over OpenAI through equity, distribution, and compute is now also building an independent deployment arm competing for the exact enterprise relationships OpenAI and Anthropic are chasing with their own forward-deployed teams. The antitrust sword has a great deal of fresh material to work with here, whenever it next swings.

Both companies are betting on the same thing, the one path that has ever let an infrastructure owner keep the fortune it built. Climb every layer at once and don't stay a pure play at any single one of them. Neither has yet proven the bet correct, and Google carries an additional exposure the other full-stack aspirant doesn't share to the same degree. It is already living under the antitrust sword: a search-monopoly case adjudicated with real remedies attached. Every additional layer Google folds into its stack—chips, models, cloud, distribution—is another data point for the next regulator who decides the pattern has repeated once too often. Owning the full stack is the identified path to keeping a fortune. It is also, on the same evidence, the most efficient way to attract the exact kind of scrutiny that eventually caps one.

The Landlord Bet: Oracle

If Google and Microsoft are trying to be GE, Oracle is the company this book's engines predict should be most exposed to the fate GE's competitors met instead: the pure infrastructure play, stranded at the bottom of the stack while the value it enables migrates past it. Oracle's numbers tell a story that has moved from theoretical exposure to a live market signal in the span of this book's drafting. Its capital expenditure ran to 83 percent of revenue in 2026, a ratio no company its size sustains comfortably for long. Its $300 billion Stargate commitment to OpenAI is the single largest bet on this book's list that depends entirely on a counterparty, OpenAI. OpenAI is running billions of dollars into loss against a revenue base still an order of magnitude smaller than the commitment it's servicing. This book reads Larry Ellison as having built Oracle's entire AI-era strategy on the wager that being the landlord to the most ambitious model maker in the industry was worth carrying more balance sheet risk than any of his hyperscaler peers had been willing to accept. Oracle's remaining performance obligations, the backlog of contracted future revenue, grew more than fourfold to north of half a trillion dollars, an extraordinary number that has so far outpaced the actual revenue showing up to fill it.

Oracle's credit rating sits at BBB according to S&P Global, two downgrades removed from junk. Its cost of credit-default protection more than tripled since September 2025, and Barclays analyst Andrew Keches has warned it could run short of cash by November 2026 without borrowing again at rates that would confirm exactly the fear the downgrades already priced in.

Compare that to the other four hyperscalers. Microsoft carries a AAA rating from both S&P and Moody's. Alphabet sits at AA+ (S&P) and Aa2 (Moody's), Meta at AA- (S&P), and Amazon at AA (S&P). All four have issued extraordinary amounts of debt in 2025 and 2026 to fund the identical AI build-out Oracle is funding, tens of billions of dollars apiece, including a hundred-year bond in Alphabet's case and the largest non-M&A high-grade corporate bond sale on record in Meta's. The difference is not the size of the debt. It's what the debt is for. For the four elite-rated hyperscalers, issuing bonds against a fortress balance sheet is production capital optimizing its own capital structure, borrowing cheaply because the borrower's fundamentals support it regardless of how the AI bet plays out.

For Oracle, the same instrument reflects the opposite. A company borrowing because the alternative is running out of runway before the bet it made resolves either way. Microsoft, notably, has structured its own AI infrastructure spending to avoid ever appearing on its own balance sheet in the first place, routing tens of billions of dollars of leverage through an off-balance-sheet joint venture with BlackRock, Global Infrastructure Partners, and the Emirati fund MGX. Oracle has no equivalent shock absorber. Every dollar of its exposure sits in full public view. Every company in this chapter is exposed to the commoditization pressures traced. Oracle is the only one where a credit rating agency has started pricing that exposure into the cost of its money.

The Arbitrage Bet: Amazon

Amazon's position was the cleanest hedge on the original map. Invest in OpenAI and Anthropic simultaneously, sell cloud capacity to both, and let Bedrock host its own Nova models alongside the rivals it's bankrolling, extracting a toll from every side of the fight without needing either side to win. The specific numbers made the symmetry almost too neat to be coincidental. A $50 billion investment in OpenAI in February 2026, followed two months later by up to $25 billion more into Anthropic on top of the $8 billion already committed, each lab separately pledging more than $100 billion in matching cloud spend back to Amazon's Trainium chips. Andy Jassy signed both deals inside the same two-month stretch, one at the end of Amazon's first fiscal quarter and the other three weeks into its second. That neutrality held through 2026 as a matter of financial structure.

It did not hold as a matter of strategic behavior. Amazon's billion-dollar commitment to forward-deployed engineering, announced two days ahead of Microsoft's Frontier Company, is Amazon abandoning pure neutrality for the identical workflow ownership bet every other full-stack aspirant in this chapter is now making. The arbitrage position and the full-stack position, treated previously as separate bets, turned out inside of a year to be a single continuum every large enough player eventually slides along. Amazon still collects a toll no matter which lab wins the model war. It has stopped being content to only collect the toll.

The Model Maker Bet: OpenAI and Anthropic

The two labs that own no infrastructure of their own, renting compute rather than building it, are exposed to a different mechanism than the hyperscalers. Not the Indispensability Trap's classic rate regulation reflex, but commoditization pressure from the model layer itself. Chapter 10 confirmed that exposure was real, tracing how quickly open-weight models converged on frontier benchmarks. How OpenAI's gpt-oss release amounted to a tacit admission that even the company most associated with keeping its weights closed no longer believed the model alone was worth defending unconditionally.

OpenAI's financial position sharpens the stakes further. A company on track to lose billions of dollars in 2026 against a revenue base still climbing toward the scale its infrastructure commitments already assume, filing confidentially for a public listing that could value it past a trillion dollars while simultaneously negotiating away a slice of that same valuation to the government reviewing the filing. Anthropic, smaller and privately held, carries a cleaner balance sheet and a correspondingly smaller infrastructure commitment, the trade-off already found running through its multicloud hedge across Amazon, Google, and now Microsoft's Copilot products.

Chapter 13 added a second exposure neither lab had fully priced into its own risk model. Pressure from above, arriving as an export control letter that could vanish a flagship product in an afternoon, or as a government invited to take a direct equity stake in exchange for political cover. Both labs are now squeezed from below by commoditization and from above by the state, a position genuinely unique to this pair on the entire map.

Their responses diverged in a way worth holding up against each other rather than treating as interchangeable risk management. OpenAI offered Washington equity directly, a bet that buying insurance against an unfavorable version of state leverage was worth the price of alignment. Anthropic proposed universal predistributive capital accounts instead, declining to hand the government a stake in the company it would eventually need to regulate. It has separately chosen to fight its own supply-chain-risk designation in federal court rather than negotiate around it the way OpenAI negotiated around the equity question.

Both labs, independent of that disagreement, launched forward-deployed engineering units of their own in May, arriving at the identical conclusion the same month Microsoft and Amazon would reach weeks later. Even the two companies that build the frontier models no longer trust the frontier model alone to hold their position. Mistral's counterexample is proof the free model bet can work as a business, a twentyfold revenue increase built entirely on the layer above a model given away for free. Neither OpenAI nor Anthropic has Mistral's clean version of that story yet. Both are hedging as if they know it.

The Hybrid Bet: xAI and SpaceX

Chapter 7 found xAI folded into SpaceX straddling both categories at once, an infrastructure owner and a model maker inside a single balance sheet, with Musk keeping Colossus 2 walled off from the leasing arrangements funding the rest of the operation. Brad Feld's rule of three, the framework I heard directly from Feld and used to explain SpaceX's posture, still holds. Grok sits outside the top three labs by most measures. SpaceX's response has been to hedge by renting out the compute a market leader would have kept for itself. Leasing capacity to Google and Anthropic, two of the very labs it competes against on the model side of its own business. Nothing in the chapters since has resolved whether that hedge is shrewd capital discipline or a tell that the model side of the bet has already been quietly written down. The lease terms, cancelable on ninety days' notice by either party, are still the most honest evidence available. Nobody involved is confident enough in this arrangement's permanence to sign anything longer.

SpaceX's path to public markets adds a wrinkle none of the other bets in this chapter carry. A single vehicle now holds a rocket company, a satellite internet business, an AI lab lagging its rivals, and a GPU-leasing operation renting to those same rivals. Additionally, it owns a coding startup that runs part of its own product on a Chinese open-weight model rather than on Grok at all. No other entity on this map has folded quite so many unrelated bets into one balance sheet. No other entity has quite so little exposed logic tying all of them together beyond Musk's conviction that owning pieces of every layer beats depending on anyone else's.

Complicated: Meta

Chapter 7 sorted Meta into its own category, disintermediation with a twist. A company giving away its Llama models for free, commoditizing the layer everyone else was trying to defend, and capturing its return instead at the social and advertising layer above it. That bet looked coherent as recently as Chapter 10, where Llama's fading benchmark position against Mistral and the Chinese open-weight wave was framed as evidence that the strategy had been overtaken by rivals playing Meta's game harder. Meta Compute complicates the story in a direction the original map didn't anticipate at all.

Initiative Infrastructure Chief Santosh Janardhan and Meta Superintelligence Labs' Daniel Gross began coleading Meta Compute in January 2026 alongside President and Vice Chairman Dina Powell McCormick. It was reported in July 2026 to be building out a cloud business selling Meta's spare AI capacity and hosted models to outside customers, the same landlord move Google, Microsoft, Amazon, and Oracle already made years ago. Mark Zuckerberg's description of the idea, that it's definitely on the table, is a tell in itself. A company that spent Part III commoditizing the model layer on purpose is now weighing whether to become a toll collector on the infrastructure layer too. Zuckerberg had previewed the logic himself months earlier, on Meta's October 2025 earnings call, noting that outside companies were already approaching Meta most weeks asking to buy compute at a premium to what Meta itself had paid for it, demand Meta had simply been declining to monetize.

If Meta Compute proceeds, Meta doesn't abandon its original bet so much as add a second one on top of it, becoming one more company chasing the full-stack strategy rather than the one confident enough in disintermediation alone to skip it entirely. Apple would be left as the only major hyperscaler-adjacent company with no public cloud arm at all, an isolation I take no position on, whether to read it as discipline or as falling behind.

Apple's absence from every chapter is its own kind of data point. The one company in this industry large enough to compete for infrastructure has been quiet. But recently with the release of their next generation Mac Mini and Mac Studio, their plan has become clearer. They are making the play to be the infrastructure on your desk, in your home, on the phone in your pocket. As the price of compute from AI factories solidifies and the power of open weight models grows, Apples own hardware, with built in neuro processors and able to work with large amounts of memory, becomes an excellent choice for running AI locally. This strategy is worth watching as it does offer several advantages, decreased cost, lower latency, more control and security, to name three.

A Map Drawn Only in American Dollars

Every bet examined so far in this chapter assumes the fight is happening inside a single American-centered map. But Chapter 11 already showed that that assumption doesn't hold globally even if it holds inside this book's borders of attention. China's answer to the question of where durable value will land is not to be playing the same game at all. DeepSeek, Alibaba's Qwen, Moonshot, and Zhipu aren't racing to own compute, chips, or deployment relationships. They're racing to become the default substrate the rest of the world builds on for free, wagering that global developer mindshare and the physical-world data advantage of China's manufacturing base matter more than owning any single layer outright.

Cursor, the startup this chapter has already traced through SpaceX's acquisition, runs on Kimi rather than an American model and is the cleanest available proof that China's bet and the American bets in this chapter are not two versions of the same contest. They're two different contests, running on separate tracks, occasionally intersecting inside the same piece of software.

Sovereign AI is what happens when a government too small to run either race decides it doesn't want to depend on the outcome of either one. The EU's gigafactories, Britain's compute fund, India's frugal domestic stack, and the Gulf states' alignment purchases aren't bets on which company wins. They're hedges against needing any of them to. That distinction matters for how to read this chapter's map correctly. Every company examined here is competing to become indispensable to as much of the world's AI activity as it can capture. Sovereign AI is the rest of the world's answer to the discomfort of watching that competition from the outside, and it is only a partial answer at best.

Which History Rhymes

Chapter 2 promised a specific discipline for testing whether this book's pattern holds a fifth time. The same six questions asked the same way of a fifth build-out unlike the four that came before it. Four protagonists carried those questions through Parts I and II, Vanderbilt, Vail, Insull, and the operator who watched the fiber glut collapse in real time. None of the bets on this chapter's map is a clean copy of any one of them. But each has a closer resemblance to one historical figure than to the other three, and naming the resemblance is the most direct test available of whether the pattern is actually repeating or whether this book has been reaching for comparisons that flatter its own thesis.

Oracle's position is Samuel Insull's, almost without needing to translate it. Insull built a holding company empire leveraged far beyond what its underlying utilities could support. He bet that growth would arrive fast enough to service debt taken on to fund growth that hadn't arrived yet. Oracle's capital structure, an 83 percent capex-to-revenue ratio and a $300 billion commitment to a counterparty still years from profitability, is the same wager in different units. Insull's empire collapsed within a matter of months once the assumption underneath it, that growth would keep outrunning the debt, stopped holding. Whether Oracle's ends the same way is the live question readers should watch most closely, and the credit rating evidence says the market is already watching it too.

Amazon and Nvidia both carry a trace of Cornelius Vanderbilt, the railroad chapter's partial exception, the man who made his fortune consolidating and financing rail lines rather than running the trains that generated the freight revenue everyone else was fighting over. Vanderbilt understood that the money was in controlling the terms other people operated under, not in operating anything himself. Amazon's toll on both OpenAI and Anthropic, collected regardless of which lab wins, and Nvidia's new revenue-sharing structure, collecting a cut of whatever its customers earn without needing to run a single data center itself, are both Vanderbilt's insight running through modern balance sheets. The railroad chapter's verdict on Vanderbilt is worth remembering exactly because it complicates the comparison rather than flattering it. Even he topped out at a fraction of what Carnegie and Rockefeller built one layer above the network he financed. Arbitrage was a strong position in 1869. It was never the strongest one available.

OpenAI's equity offer is Theodore Vail's bargain updated. Vail secured the Bell System's survival by negotiating directly with a government that could have broken the company apart, trading a measure of independence for a settlement on terms Vail helped write rather than terms imposed on him later. OpenAI's proposed 5 percent, whatever else it is, is the identical move. Negotiate the terms of eventual state involvement before the state decides those terms without you in the room. Vail's bargain held for the better part of a century. Whether OpenAI's holds for even a fraction of that is a question that is genuinely unresolved.

The internet chapter's operator, the witness to the fiber glut who watched vendor financing dressed up as organic demand right up until the demand failed to arrive, has the least comfortable modern echo of all. That position belongs to every company in Chapter 8's circular financing web at once, and to the sixty-day forward-deployed engineering convergence that opened this chapter. A frenzy looks identical to genuine demand for exactly as long as nobody needs the money to come from somewhere else. This book's fifth build-out has that same tell running through nearly every bet examined. The four historical precedents don't just rhyme with today's map. Between them, they cover almost the entire map.

No Safe Rung, Confirmed

Lay all these bets next to each other and the pattern stops being a hypothesis and starts looking like a simple description of what actually happened. Every company on this map, without exception, is currently doing something it would not have needed to do if the layer it originally bet on had turned out to be defensible on its own. Google and Microsoft are climbing every layer because no single layer—not chips, not cloud, not the model—held enough value alone to bet the company on. Oracle is exposed precisely because it tried to be a pure infrastructure play in an era that punishes pure infrastructure plays first. Amazon abandoned strict neutrality once neutrality alone stopped looking like it was enough. OpenAI and Anthropic are hedging against a government that didn't exist as a variable in their business plans eighteen months ago. Meta is reconsidering the one bet that looked most confidently different from everyone else's. Nvidia, the company with the strongest claim to indispensability of anyone on this list, chose to become a bank rather than trust that indispensability to hold its value unassisted. SpaceX folded a fourth business into an already improbable conglomerate rather than let its model business stand or fall on its own.

This is the Indispensability Trap operating exactly as described, only faster and across more layers at once than any of the four historical build-outs managed. The railroad chapter needed decades for rate regulation to arrive after the abuse it was responding to. The telephone chapter needed a hostile Justice Department circling before Vail's universal service bargain made sense as a defensive move. This build-out has compressed that same defensive reflex into months and distributed it across every layer of the stack simultaneously rather than letting each layer take its turn.

Jevons paradox explains why the race keeps accelerating rather than settling into something calmer. Cheaper models and cheaper compute don't shrink the prize. They expand the volume of AI running through the economy, which raises the stakes on whichever layer eventually proves defensible rather than lowering them. Every efficiency gain documented—DeepSeek's training breakthrough, the inference cost collapse, Nvidia's chip roadmap—has been followed by more capital chasing the newly cheapened opportunity, not less.

The Perez framework's installation frenzy is the honest name for the sixty-day convergence on forward-deployed engineering that opened this chapter, capital and headcount racing toward an unproven destination because standing still looks like the only certainly wrong answer available to any company this size. The distinction between financial and production capital is the one Oracle's credit rating has already started to answer in public in a way the other bets on this map have not yet been forced to confront. Whether the rest of this book's map is quietly running the same experiment at a slower pace is the question every one of these companies is being asked by its own shareholders right now, regardless of whether the shareholders are using this book's language to ask it.

None of this hands a verdict to any single company on the list, and the honest version of this chapter has to end without giving you one. Google's distribution advantage is real and, on the evidence gathered here, the least contested position on the entire map. Microsoft's hedge is rational, and its stock price is skeptical of it anyway, a genuine disagreement between the company's strategic logic and the market pricing that logic in real time. Oracle's exposure is the most acute on the map, and it is also the clearest bet on a genuine, still-unresolved wager, that OpenAI's revenue eventually catches the commitment Oracle made against it, at which point the same balance sheet that looks reckless today looks prescient.

Amazon and Meta have both quietly abandoned the cleaner, more differentiated bets Chapter 7 originally assigned them, converging toward the same full-stack strategy Google pioneered. This either validates Google's read of the market or simply means every company big enough to try is now trying the identical thing at once. Chapter 1's opening argued this and it is itself a warning sign rather than a reassurance. OpenAI and Anthropic remain the two participants most exposed to forces outside their own control, squeezed by commoditization from below and the state from above, with no equivalent in the four historical build-outs for a company facing both pressures at once this early in its life. Nvidia's shadow-bank pivot is either the shrewdest hedge in this entire book, a company converting its temporary chip advantage into a permanent financial one before the chip advantage itself erodes, or a new layer of hidden leverage nobody has stress-tested yet, dressed in language borrowed from Silicon Valley rather than Wall Street.

A reader checking back on this chapter in a year has sharper things to watch for than a general sense of who looked strong. Oracle's numbers will say plainly whether the Insull comparison held. A further ratings downgrade or a forced renegotiation of the OpenAI commitment would confirm it, while revenue actually catching up to the backlog would be the clearest possible refutation this book's engines could ask for. The Microsoft-OpenAI FTC investigation resolving, in either direction, will say whether the antitrust sword has a sixth target after Google or whether the structure survives the same scrutiny that caught IBM and Microsoft's earlier incarnation. Whether Meta Compute actually launches, rather than staying something Zuckerberg described as merely on the table, will say whether Meta's original disintermediation bet was a considered strategy or a position it could only afford to hold before the rest of the industry proved the alternative out. Nvidia's repurchase guarantees will eventually be tested by an actual default somewhere in its partner network, at which point the market finds out in one transaction whether the shadow-bank bet was well-collateralized or merely well-marketed.

What every one of these bets shares is the thing the General Electric case predicted from the very start of this book. Durable value keeps migrating up and across the stack faster than any single company can climb after it. The only players with any real chance of keeping what they've built are the ones willing to keep climbing, indefinitely, with no guarantee the climbing ever ends at a floor that holds. GE's early-twentieth-century advantage was that it had a full human generation to make that climb before its competitors figured out the same trick. Every company examined in this chapter is making the identical climb in real time, in public, with its shareholders watching the ascent and pricing the risk of a fall before anyone can be sure the summit exists.

Part III opened by asking whether owning the full stack finally lets an infrastructure owner keep what it built. Seven chapters later, the honest answer is that everyone capable of trying is trying, all at once. The state itself has become a new kind of participant in the outcome, deciding, sometimes with a letter and sometimes with a stock certificate, who gets to keep playing at all. Where this settles, whether any of these bets resolves into something as durable as General Electric once did or scatters the way the pure-play railroads and telephone carriers did before it, is a question nobody can answer from inside 2026. What Part IV turns to next is a different question entirely, and in some ways a harder one. Not who wins the fight over value inside this stack, but what it costs everyone standing outside it while the fight continues.

Part Four: The Stakes

Why regulation and safety matter.

Chapter 16

The AI Industrial Complex

> In the councils of government, we must guard against the acquisition of unwarranted influence, whether sought or unsought, by the military-industrial complex. The potential for the disastrous rise of misplaced power exists and will persist.
>
> —President Dwight D. Eisenhower, Farewell Address, January 17, 1961

In May 2026, Google's former chief executive turned elder statesman, Eric Schmidt, stood up to deliver a commencement address and got booed. It happened at the University of Arizona. A similar scene had played out a week earlier at the University of Central Florida, where a different speaker, real estate executive Gloria Caulfield, was booed after calling AI "the next industrial revolution." The graduates doing the booing at both schools were not, by any reasonable reading, anti-technology. Most of them had grown up with a smartphone in one hand and a search bar open in the other. What they were reacting to was something more specific. Every prior wave of consumer technology examined, the personal computer, the internet, the smartphone, arrived selling itself as empowerment, a tool that made an individual more capable. AI is the first wave that has, at least in its loudest public voices, sold itself as a replacement instead.

To be fair, the college graduates booing at graduations are facing one of the toughest job markets seen in some time. What makes it worse is that the usual advantage of a college degree has narrowed to its thinnest margin in decades. Recent graduates still have a lower unemployment rate than young workers without a degree, but for the first time on record their unemployment rate has climbed above the national average. For right or wrong, all the commentary about AI taking your job, specifically that AI will replace junior-level workers, has struck a raw nerve. Many of these recent grads has the double indignation of also facing student loans, typically in the $20,000 to $40,000 range for a bachelor's degree but often climbing into six figures for those who went on to graduate school, that need to be repaid. It's tough to repay a loan when you don't have a job.

The resentment has a second, harder edge underneath the generational one, and it shows up in a pattern the next chapter examines in full. Publicly traded companies simultaneously announcing layoffs and record AI infrastructure spending, often in the same quarterly earnings call. The causal link between any single layoff and any single AI deployment is rarely as direct as the headlines suggest. But the optics are direct regardless, and optics are what erode legitimacy fastest.

That erosion looks sharper set against earlier chapters than it would in isolation. Part II spent four build-outs profiling men whose fortunes rivaled anything AI has produced so far, and nearly all of them eventually reinvested some share of that fortune into the public sphere they'd extracted it from. Rockefeller funded the public health and research institutions that still bear his name. Carnegie built more free public libraries than any private citizen before or since. Henry Ford paid wages high enough, most famously the 1914 Five Dollar Day, that popular legend still credits it with letting his own workers afford the cars they built, even though Ford's internal motive was mainly to curb ruinous labor turnover. Even Bill Gates, a generation closer to the AI chapter than to the railroads, built one of the largest philanthropic foundations in human history.

None of this makes any of them saints, and this book has not treated them as such. It does mean each of them, eventually, made some visible gesture toward the society their fortune was extracted from. The AI era's wealthiest figures have, so far, offered something closer to the opposite impression. Transactional, competitive, and largely dismissive of the public unease this chapter opened with. Saying it even plainer, the group of so-called "tech bro" billionaires who are the faces of the AI movement are not a likable bunch and seem to care for no one but themselves. Whether that impression is fair to any single individual is beside the point. The impression is the thing eroding trust, and trust is the resource this chapter is about.

That reality is something to be acknowledged before this book's final part goes looking for who bears the cost of everything Parts I through III have described. Whether the resentment is fully justified or only partly is beside the point. The reaction is real, it is generational, and it is aimed squarely at a small number of people who have accumulated more simultaneous influence over compute and capital and now government policy than any comparable group in any of the four build-outs that came before.

An Old Warning, Updated

Dwight Eisenhower delivered his farewell address on January 17, 1961, three days before leaving office, in a speech that took less than ten minutes to deliver and has outlived nearly everything else said in American politics that year. Eisenhower was not a pacifist making an abstract point. He was a five-star general who had run the Allied invasion of Europe warning the country he had just finished defending that the alliance between the Pentagon and the arms industry that had won the war now risked outliving the war's necessity, concentrating power in a way no democracy had previously had to guard against. He named it plainly: the military-industrial complex.

I was a small child during Vietnam, too young to understand the war itself but old enough to absorb the fact that the phrase carried weight every time an adult said it on the news. It took years to understand why. Eisenhower's credibility was the point. A career military officer telling Americans to distrust the alliance between the military and the industry that supplied it was not an activist making noise from outside the system. It was the most credible person inside the system saying the system itself had grown a blind spot.

Sixty-four years later, on his own way out the door, Joe Biden reached for the identical warning and updated it for a different industry. He told Americans he was concerned about the potential rise of a tech-industrial complex that could pose real dangers for the country. Five days after that, the new administration was sworn in with Mark Zuckerberg, Sundar Pichai, Jeff Bezos, Elon Musk, and Tim Cook seated more prominently at the inauguration than the incoming cabinet nominees. Whatever else that arrangement was meant to communicate, it communicated this: The same handful of companies mapped across Part III, the ones fighting each other over infrastructure, chips, models, and now forward-deployed engineers, are also, simultaneously, standing closer to the seat of government than the people about to run its agencies.

Eisenhower's speech contained a second warning that gets quoted far less often than the first, and it may be the more precise fit for what this chapter is describing. In the same address, he cautioned against the domination of national policy by what he called a scientific-technological elite, a class of researchers, administrators, and funders whose command of specialized knowledge let them shape public decisions that elected officials no longer fully understood well enough to challenge. Eisenhower worried about physicists and engineers occupying that role in 1961. In 2026 the people occupying it build and sell frontier AI models, and the government agencies meant to oversee them increasingly lack the in-house technical expertise to evaluate what they're being told. A regulator cannot meaningfully discipline a system it does not understand well enough to audit, and the companies building these systems are, for now, the only parties who understand them fully.

The Mechanism, Restated

What makes this book's argument more than a coincidence of timing is that the mechanism connecting these companies to government isn't new, and it isn't unique to this administration. It's the same reinforcing loop discussed since the start, the Indispensability Trap, working exactly as designed. Demand for advanced computing and AI capability drives investment. Investment produces increasingly powerful systems. Those systems, once deployed inside government agencies, businesses, and critical infrastructure, become difficult to replace, creating a dependency that didn't exist when the contract was signed. The companies supplying that dependency stop being vendors and start being what both industry and government now call strategic partners, a title that comes with privileges no ordinary contractor receives, and with an expectation of continued access that ordinary contractors never assume either.

This isn't a hypothetical mechanism. It's the same one found at the silicon layer, when the state took an equity stake in Intel rather than wait for a crisis to force the question, precisely because a chip shortage discovered after the fact would already be too late to fix. It's the same one found at the model layer, when an export control letter could suspend a frontier model worldwide in an afternoon. When a lab facing that risk chose to offer the government equity before the government asked for it, calculating that a voluntary stake was cheaper insurance than an imposed one.

What this chapter adds is the recognition that this loop isn't confined to any single layer of the stack. It runs through the entire relationship between the companies building AI and the government whose policy increasingly depends on them, and it was running before this book gave it a name. Eisenhower himself named the same loop in miniature in his own farewell address when he described a government contract becoming virtually a substitute for intellectual curiosity. The money doesn't just fund the research anymore. It starts to shape what gets researched, and by whom, and for whose benefit the answer eventually gets used.

The clearest single face of it belongs to Palantir, first as the originator of the forward-deployed engineering model that Microsoft, Amazon, OpenAI, and Anthropic all converged on inside the same sixty-day window, and again in its classified systems partnership with Nvidia running open models inside air-gapped government deployments.

Palantir's chief executive, Alex Karp, has not confined himself to selling software. He has taken to describing what counts as patriotism and what doesn't in public commentary that positions Palantir not merely as a vendor to the national security state but as a voice helping define its values. That is a different order of influence than Microsoft holding an equity stake in OpenAI or Google training its own frontier model. It is a private company's chief executive presuming to arbitrate the meaning of a civic virtue on behalf of the government paying the company's invoices. The original military-industrial complex built the tools of war. The emerging technical version is building the systems that help determine how those tools get used, and, in Karp's case, the ideas that justify using them at all.

What makes Karp's position more than an isolated eccentricity is how quickly the practice he pioneered became the industry's default. All four companies previously mentioned have adopted the exact same operating posture, engineers embedded inside the client's walls, that made Palantir's deeper influence possible in the first place. The business model spread so fast that no one noticed the values attached to it were spreading too.

The Layers, Mapped

The ecosystem this produces isn't flat, and it's worth laying out in the same layered terms Part III used to map the companies fighting over infrastructure. At the base sit the hyperscalers, Microsoft, Amazon, Google, whose cloud contracts increasingly double as the plumbing for federal agencies as much as for private enterprise. A government agency migrating its case management or its intelligence analysis onto Azure or AWS is entering the identical dependency relationship described via chipmakers entering with a single foundry. One layer up sits what amounts to the digital environment itself, Google's search index, Meta's social graph, Apple's device ecosystem. Platforms so embedded in daily civic life that a change to any one of them functions closer to infrastructure policy than product strategy.

Nearly four in ten American adults get news from Facebook, and more than a third get it from YouTube, on recommendation algorithms that no regulator outside the company itself can fully audit.

A third layer, newer and stranger than either of the first two, belongs to companies whose infrastructure has begun operating at a scale usually reserved for nation-states. SpaceX's Starlink network, which by 2026 had become consequential enough to the war in Ukraine that decisions made inside a single private company, which regions retained connectivity and which did not, carried the weight of a foreign policy choice no elected official had made.

At the newest, least settled layer sit the frontier AI labs themselves, OpenAI and Anthropic chief among them, building decision support systems that are, by design and by necessity, harder to audit than anything in the three layers beneath them, precisely because the systems are meant to reason rather than merely execute.

None of these four layers was built with the others in mind, and none of the companies inside them set out to become a segment of anything resembling a complex. That's precisely Eisenhower's point restated at AI's scale. The danger was never a conspiracy. It was aligned incentives, pursued honestly by everyone involved, that happen to produce a concentration of influence no single actor intended and no existing institution was built to check.

How Far the Infrastructure Goes

The clearest physical evidence of how deep this dependency now runs isn't in Washington at all. It's on the ocean floor and, increasingly, in orbit. In 2026 an AI data center began operating underwater off China's coast, powered by offshore wind turbines and cooled directly by the surrounding seawater rather than the increasingly contested fresh water the next chapter examines. The engineering logic is straightforward once stated. Seawater at depth stays cold year-round without mechanical refrigeration, and a facility with no human occupants to protect can tolerate pressure and corrosion trade-offs no office building ever could. Around the same period, SpaceX began advancing serious plans for AI compute infrastructure in orbit, solar powered around the clock without atmosphere or weather to interrupt collection and cooled by radiating heat directly into the vacuum of space rather than the water tables and power grids straining under every terrestrial data center. Neither project is really about the ocean or about space. Both are evidence that land-based infrastructure, power, cooling, water, permitting, and the interconnection queues that are already straining under the silicon and electrification build-outs, has become constrained enough that going to the bottom of the ocean or the edge of orbit now reads as a rational engineering response rather than a publicity stunt.

I keep coming back to electrification as the closest comparison for what's happening here, closer even than the internet's build-out. The specific application, the light bulb, the motor, the industrial process running on the current, mattered far less in the end than the fact of the grid itself, universal, taken for granted, invisible the moment it worked. AI's infrastructure is following an identical path but at a pace that compresses decades of the electrification chapter's timeline into a handful of years. The ocean floor and low orbit are simply where the grid goes once every accessible acre of dry land has already been spoken for.

That raises a policy question we have been circling since the silicon chapter and sovereignty chapter. Should infrastructure this fundamental, physically extending past the edges of any single nation's jurisdiction, be guided primarily by the commercial decisions of a small number of corporations answering to their own shareholders? Can any single government credibly claim to manage something that operates on the seabed and in orbit simultaneously when the same government's agencies are, per this chapter's opening, increasingly dependent on the companies it would need to regulate?

Europe and parts of Asia have at least attempted an institutional answer. Several European labor frameworks give organized labor a formal seat at the table when automation decisions get made, not a veto but a documented right to be consulted before a workforce gets restructured around a new system, a right American workers have no equivalent claim to. South Korea has, at points, seen its largest chipmakers negotiate formal bonus pools, government mediated and tied to a fixed share of operating profit, that link AI-driven productivity gains to a negotiated payout rather than leaving the distribution of those gains to each company's discretion alone. Neither approach has fully solved the problem this chapter describes, and I take no position on which national model gets the balance right. What both demonstrate is that the absence of any equivalent American institution is a choice, not an inevitability, the same way universal service was a phrase Theodore Vail coined. A bargain the country negotiated over decades rather than a default outcome of the technology itself.

The US, so far, has largely left the question to the companies building the infrastructure and the market pricing their bonds. Nobody currently running for office has a complete answer either way. The scale of the build-out has outrun the institutions meant to govern it, the same gap found at the silicon layer and at the model layer, now visible at the level of the physical planet itself. A government agency cannot audit a system its own staff cannot fully explain, and the fastest-growing item in that agency's budget is frequently the contract with the company whose system it would need to audit.

What This Chapter Is For

On July 20, 1969, Neil Armstrong described one small step for a man becoming a giant leap for mankind, a line he later said evolved in his own mind in the hours before landing to capture the strange scale of the moment. An act small enough for one body to perform yet consequential enough to belong to an entire species. The infrastructure this chapter has described is being built by a small number of men, and a handful of women, whose names have filled this book's chapters since Part III began, and it invites the same line read a different way.

It is entirely possible to build something this consequential as one small step for the men who happen to own it, extracting the gains for a boardroom and a shareholder base, rather than the giant leap for mankind the original moon landing at least aspired to be. Nothing in the physics of an underwater data center or an orbital compute cluster determines which version of that sentence turns out to be true. That is decided by decisions Part III described as bets made by boardrooms accountable mostly to their own shareholders in the absence of any institution positioned to ask the question on behalf of everyone else.

The chapters that follow go looking for the people standing outside the map this book has spent Part III drawing. They are the communities whose water and land and quiet get spent on a data center they were never fully told the purpose of, the workers whose jobs get reallocated to fund a bet that hasn't paid off yet, the publishers and creators whose work trains the systems now competing with them for the same audience. The citizens whose government has begun depending on the companies it's supposed to hold accountable. Eisenhower's warning was never that the alliance he named was evil. It was that an alert and knowledgeable citizenry was the only check available against its unwarranted influence, and that citizens who stopped paying attention would eventually find the choice had already been made without them. Sixty-five years later, that's still the only check we can offer. What's changed is how much faster, and how much further from land, the choice is now being made.

Chapter 17

Who Bears the Cost

> The cost of a thing is the amount of what I will call life which is required to be exchanged for it.
>
> —Henry David Thoreau, *Walden*, 1854

The trees along South Pine Street in Spartanburg County, South Carolina, came down before most of the county knew what was replacing them. For months, county officials and the developer referred to the project only as Project MOC-1, a code name attached to a description, a "high-performance computing center," vague enough to mean almost anything. Even after the developer's identity became public, officials continued for over a year to avoid the words "data center." The site itself had history. It was the former home of Kohler's vitreous works, a plant that spent sixty-six years turning clay and fire into sinks and toilets, the ordinary fixtures of an American house, before closing in 2024. By the time the public learned that MOC-1 was, in fact, a $2.8 billion data center backed by a New York investment firm called NorthMark Strategies operating through a subsidiary named Valara Holdings, the land was already cleared and construction equipment was already active on the site.

This wasn't a failure of communication. County Councilman David Britt, the official who first described the project publicly as a computing center rather than a data center, later took a position as the state's public service commissioner, where he continued defending the project and was later the target of a recusal motion filed by opponents who said his advocacy made him unable to rule on it impartially. Nancy Garner, who lives near the site, put the result in the plainest terms available. "I feel totally deceived by the county council, by anyone that supported that data center in the middle of our neighborhood. I feel like they couldn't care less about our property, our water, our air."

Spartanburg County offered NorthMark a property tax rate of 4 percent for forty years, well below the 6 percent standard rate manufacturers otherwise pay, plus credits worth up to $2 million a year, structured so their total value could never exceed NorthMark's investment. County Councilman Britt defended the arrangement at the time by noting that unemployment in the county already sat at 3.9 percent, below the state average of 4.7 percent, meaning the twenty-seven jobs on offer were never really pitched as solving a local unemployment problem so much as adding tax revenue on top of an economy that was, by the county's numbers, already doing fine.

In exchange the county was promised a facility that would generate its own power from an adjacent natural gas pipeline rather than straining the local grid, consuming no more energy than Kohler's vitreous plant once had, and creating those twenty-seven full-time jobs, a number for which NorthMark has never disclosed salary or role details. Months after construction began, Valara Holdings asked South Carolina's environmental regulators for permission to add additional gas-fired turbines beyond what was originally described. State filings show the request would push the facility's output from roughly 50 megawatts to 450 megawatts and make it one of the county's largest emitters of formaldehyde and fine particulate matter. The limits, the Southern Environmental Law Center argues, are written loosely enough to be functionally unenforceable. Self-generation was offered to residents as a reassurance that the project would sit lightly on the county. The reassurance turned out to be the leading-edge of the concealment, not its resolution.

Spartanburg is one of roughly three thousand data centers now operating in the United States, with another fifteen hundred under development. It may be the starkest illustration of a pattern the previous chapter named but didn't yet prove. Namely that the same aligned incentives Eisenhower warned about, pursued honestly by everyone involved, can produce outcomes that look, from the outside, indistinguishable from a conspiracy. Nobody in Spartanburg conspired to deceive the public in the legal sense. Local officials wanted the investment and the tax base it promised. The developer wanted a permitting process free of the delays public scrutiny tends to invite. Both parties got exactly what they wanted, and the residents living along South Pine Street got a fait accompli.

The Math That Doesn't Work

Start with the number planted all the way back in Chapter 1: the capital-to-jobs ratio, capital on the scale of the steel age, employment on the scale of a corner store. Nowhere has that ratio been tested more directly than in the small towns now hosting this build-out's physical infrastructure. The national data behind Spartanburg's experience makes the pattern impossible to dismiss as an isolated bad actor. A Michigan data center, built by Switch Inc., received tax incentives in 2015 after promising a thousand jobs within a decade. By 2022 the company had hired twenty-six people, at an average wage below $38,000 a year. Virginia's economic development data puts the statewide ratio at one permanent job for every $13 million invested. A Microsoft facility in Illinois received more than $38 million in state sales tax exemptions in exchange for twenty permanent jobs. None of these numbers describes fraud. Each describes a contract that got exactly what it promised on paper, while promising far less than what most residents assumed they were going to get.

Broadview, Montana, a town of roughly 150 people north of Billings, is watching this ratio arrive in real time. A company called Quantica Infrastructure has secured a five-thousand-acre parcel south of town, thirty-one times the size of Broadview itself, for a project it calls Big Sky Digital Infrastructure. Public filings with the local utility describe interconnection requests scaling toward more than seven thousand megawatts across multiple phases, an amount that would exceed the output of the Grand Coulee Dam, the largest power station in the country. Data centers in Montana are taxed at nine-tenths of one percent of market value, a lower rate than an average homeowner pays and lower still than what any other commercial or industrial land in the state is assessed.

Quantica does not currently hold water rights and has proposed a mix of zero-water air cooling, deep aquifer wells, and treated greywater without yet committing to which combination it will use once the environmental assessments are complete. Depending on the design chosen, a facility this size could draw anywhere from three hundred thousand to five million gallons of water a day, a range wide enough that residents have little basis to plan around whichever end of it Quantica eventually settles on. Anne Hedges, who has worked at Montana's leading environmental watchdog group for more than three decades and now serves as its executive director, put the calculation she believes is driving developers to the state plainly. These companies might be looking for "easy pickings" in Montana.

The people raising questions about the gap in the amount of water needed for the data center are not, based on the evidence, cranks or professional agitators. Tim Wipf, who leads a Hutterite farming colony near Broadview whose irrigation depends on the same aquifer the project may eventually draw from, put the frustration in the plainest terms available: "These people do think we are stupid." Cari Olson and Kassi Solberg, both residents, have spent months trying to get a special zoning ordinance passed in a county with no zoning at all for that stretch of land, facing what one independent account of the fight described as "official silence, procedural barriers, and open hostility from some local officials." Quantica's lobbyist offered a different diagnosis, telling a public hearing that much of the local apprehension stemmed from a general fear of artificial intelligence rather than from anything specific to this project. Whether that's a fair read of the room or a convenient way to dismiss it is a question we will leave to you the reader. What isn't in dispute is the asymmetry: A five-thousand-acre, multigigawatt commitment, evaluated by a county government with no zoning authority to condition it on anything at all.

Twenty-Three Atom Bombs a Day

Box Elder County, Utah, offers the same pattern at a scale that makes Broadview look modest by comparison. The Stratos Project, backed by the investor Kevin O'Leary, of *Shark Tank* fame, through a company called O'Leary Digital, working alongside Utah's Military Installation Development Authority, has been approved to build across roughly forty thousand acres of high desert near the Great Salt Lake, an area larger than Bryce Canyon National Park. At full build-out the campus would draw up to nine gigawatts of power, more than twice what the entire state of Utah currently uses, generated largely on-site from natural gas rather than drawn from the existing grid. State boosters describe the project as a chance to fund modernization at nearby Hill Air Force Base, create two thousand high-paying jobs, and keep the United States ahead of China in the AI race. If that sounds familiar, it should. It's the same national security framing the last chapter found running underneath the entire industry's relationship with government.

A Utah State University physics professor named Rob Davies ran the numbers on what nine gigawatts of gas-fired power generation actually means for the valley it would sit in, and no paraphrase captures the scale of what he found. The project's waste heat, once electricity generation losses and the servers' own heat output are accounted for, would pour thermal energy into Hansel Valley at a rate of roughly sixteen gigawatts around the clock, an output Davies compared to twenty-three atom bombs' worth of energy dumped into the local environment every single day, indefinitely. His own modeling suggests nighttime temperatures in the valley could rise eight to twelve degrees Fahrenheit. The project sits beside the Great Salt Lake, which is already approaching record-low water levels after a dry winter. Developers have secured roughly three thousand acre-feet of water rights for a closed-loop cooling system whose actual draw, one company executive insisted publicly, would be a fraction of what critics have claimed.

One environmental economist involved in reviewing the project raised the same question asked of Oracle's balance sheet in Chapter 15. A data center this size represents a bet that the underlying demand will still be there decades from now, and if climate pressures or a slowdown in AI investment arrive first, the risk of a stranded asset, a nine-gigawatt commitment nobody needs anymore, sits with the region that hosted it rather than the investors who financed it. Utah's approval process moved fast enough that the county commissioners voted to approve the project on May 4, 2026, before most of the reporting this chapter draws on had even been published. Developers had put roughly $20 million into a project ultimately expected to cost more than a hundred billion. That ratio, twenty million committed against a hundred billion projected, is itself a small illustration of Chapter 8's installation frenzy argument. The commitment that clears a county vote is a tiny fraction of the capital the vote actually authorizes.

What the Data Centers Don't Pay For

Threaded through all three sites is the same unresolved question about who eventually pays for the electricity, water, and land these projects consume, and the industry's preferred answer, so far, has been everyone except the industry. Montana's ratepayers, per the state's environmental advocates, may see their bills rise to fund the substations and transmission lines new data centers require, infrastructure that stays on the utility's books for decades even if the data center itself operates for a fraction of that time or shutters entirely once whichever company was leasing the capacity moves on to a newer facility elsewhere. NorthWestern Energy's Montana customers have already absorbed a roughly 22 percent increase in monthly bills since 2020, a rise that predates most of the data center build-out now under discussion and gives residents a preview of where the costs of the next wave are likely to land. This is the institutional gap restated in the most literal terms yet. European labor frameworks at least give workers formal standing when automation restructures their jobs. Nothing comparable exists for American communities asked to absorb the physical footprint of the build-out that funds it, and the absence is a choice rather than an inevitability.

Not a Tech Story Anymore

There's a fourth cost this chapter hasn't yet named, and it belongs less to any single community than to the structure of the build-out itself. Who actually owns the site once the ribbon gets cut, and what that ownership implies about what kind of asset Part III has been describing. Abilene, Texas, hosts the flagship campus of Stargate, the mega project announced at the White House by the President and leading administration officials along with SoftBank, Oracle, OpenAI, and other tech companies. The Abilene facility, just the first in the plans, is an OpenAI-Oracle infrastructure venture built on the Stargate commitment.

The site's ownership structure is worth walking through slowly because it looks nothing like the story most coverage of AI infrastructure tells. The land is owned by Lancium. The campus was developed by Crusoe, an energy and compute company, in partnership with Oracle. The roughly $9.6 billion in construction financing came almost entirely from JPMorgan, arranged by Newmark, a commercial real estate brokerage, the same kind of firm that arranges financing for shopping malls and office towers. Blue Owl Capital, an alternative asset manager, and Crusoe itself put in another $5 billion in cash equity. Oracle and Microsoft are, in the most literal sense the word allows, tenants. Oracle has signed a fifteen-year lease; Microsoft a twenty-year one. Neither company owns the building it's running its models out of.

That structure is not a one-off financing quirk. It's the emerging default. In October 2025 a consortium anchored by BlackRock's Global Infrastructure

Partners, joined by the Abu Dhabi fund MGX and a newly formed vehicle called the AI Infrastructure Partnership, whose own strategic participants include Microsoft and Nvidia, agreed to acquire Aligned Data Centers, a fifty-campus, five-gigawatt platform across the Americas, for roughly $40 billion. The sellers were Macquarie Asset Management, another large infrastructure investor, meaning Aligned had already been owned and priced as an institutional infrastructure asset once before this deal even happened. The new consortium's financial anchors include the Kuwait Investment Authority and Temasek, joining MGX in the same category of sovereign wealth capital Chapter 12 found buying privileged access to the American stack. Though here it is Abu Dhabi's MGX, not Kuwait or Singapore's funds, that maps onto the national champion pattern. BlackRock's chief executive, Larry Fink, described the ambition in the same register you'd use to describe any other infrastructure fund, saying the partnership was "mobilizing private capital to accelerate AI innovation and drive global economic growth and productivity," with a stated target of $30 billion in equity and the potential to reach a hundred billion once debt is added. BlackRock's public materials now describe data centers in the same sentence as pipelines and railroads, distinct asset classes in a diversified infrastructure portfolio, not technology bets at all.

What this means is that the deep pockets of the hyperscalers drawing on their own cash reserves and ongoing revenue to fund this build-out are no longer sufficient on their own (as Chapter 8's circular financing web already showed) and haven't been for some time. Oracle's finances make the case starkly. A planned expansion of the Abilene campus, which would have pushed it from roughly 1.2 gigawatts toward 2, was quietly scrapped in early 2026 after OpenAI and Oracle failed to reach terms. At around the same period, reporting described Oracle carrying more than a $100 billion in debt against negative free cash flow, the exact exposure Chapter 15 traced through Oracle's credit rating. OpenAI's compute chief said publicly that the company had considered expanding the site further and chose to direct that capacity elsewhere instead. A small, concrete instance of the same company simultaneously offering the federal government equity in itself while quietly redirecting its own infrastructure ambitions because the numbers on a single expansion didn't close.

None of this makes the AI build-out a fraud or a bubble on its own terms any more than a commercial mortgage makes an office tower a fraud. It does mean the honest description of what's happening on the ground in Abilene, in the BlackRock-Aligned deal, and increasingly across this entire industry, is closer to commercial real estate development than to a technology company building its own future. The landowner, the developer, the lender, and the tenant are four different parties with four different balance sheets and four different exit strategies, exactly the structure Chapter 2 found separating the grid from the appliance in every prior build-out, except here the separation runs one layer deeper than previously traced. Not grid versus appliance, but the physical building itself versus the company whose name ends up in the headlines when it opens. The pattern, though, has all the makings of ending the way other build-outs financed by other people's money have ended, with bankruptcy and someone holding the empty bag. But this time it could be even worse, as we will discuss next.

Who Bears the Words

The costs traced so far are physical. Land, water, heat, and the tax base a promised job never quite delivers. A second kind of cost runs through the information economy as well. It's easiest to see in a company Chapter 15 already mapped as the least contested position on this book's entire board.

Google AI Overviews answer a search query directly in the search results themselves, drawing on the content, reporting, analysis, and expertise of publishers and creators who built the open web. The user gets an answer. The publisher whose work the answer was built from gets nothing, because there was no click to monetize in this new reality. This is the mechanism behind what the industry has started calling zero-click search. Overall query volume keeps growing, sometimes accelerating, while the share of those queries that ever sends a visitor onward to the source collapses at the same time. Both trends are true simultaneously, which is exactly what makes the pattern easy for a platform to describe as growth and easy for a publisher to describe as extraction from the same set of numbers.

Google didn't win the fight for search against ChatGPT or any other rival lab. It won by embedding AI into a search habit it already owned and has been judged a monopolist for. That is precisely the distribution advantage Chapter 15 found running through Google's entire full-stack bet. The same advantage that makes Google's position so hard to dislodge is what makes the position of everyone whose content trains and feeds that advantage progressively weaker.

This isn't a hypothetical concern for the me. Techstrong, the media network I run, earns its living covering the tech industry. Google is building the system now competing with publishers for the same audience's attention, and I have watched referral traffic behave exactly the way this dynamic predicts. Search volume holds steady or grows while the traffic that volume used to send to the original source erodes underneath it.

The lesson isn't that publishers should give up on search-driven traffic entirely. It's that a business model built entirely on being someone else's free input, the same trap Chapter 2 described the appliance layer avoiding by building something the grid owner couldn't simply absorb, is no longer a safe one to run. The publishers surviving this transition are the ones building direct relationships instead. Subscriptions, membership communities, proprietary research nobody else can generate, the same logic Chapter 10 found driving Mistral's business one layer above a free model. The ones that don't are discovering, later than they should have, that they were never the audience Google's product was built to serve. They were an input to it.

Losing Jobs for AI, Not to AI

The clearest single data point on where Part III's bets are costing real people something concrete arrived in a wave of layoffs that ran alongside record AI capital expenditure through 2025 and 2026. Well over a hundred thousand tech jobs were cut in 2025 alone. Some broader trackers reported the actual number as approaching a quarter million, with another ninety-six thousand in the first months of 2026. The layoffs and the spending are not, in most cases, the same decision wearing two different names. Meta cut roughly 10 percent of its workforce while ramping its AI infrastructure spending. Microsoft offered buyouts to longtime employees while continuing the capital expenditure that produced its own worst stock performance among mega-cap technology companies, the same slide Chapter 15 found running alongside Frontier Company's launch. Oracle cut thousands of roles and absorbed billions in restructuring costs while its own executives described AI investment as a competitive necessity rather than a choice, the same posture that left Oracle's credit rating exposed. Salesforce's chief executive was more explicit than most, telling a podcast interviewer in plain terms that the company needed fewer heads because AI now handled work those heads used to do, a rare public admission that made the causal link direct rather than merely coincidental.

The honest description of what's happening in most of these cases, though, is capital reallocation, not direct automation. The jobs being cut are frequently not the jobs AI has actually learned to do yet. They're the jobs whose budget got redirected toward the bet that AI eventually will do those jobs, a wager on the installation-phase frenzy, drawing on the same shrinking pool of real revenue Chapter 8 identified, now claimed twice over by the layoffs tracked in the previous paragraph. The organizations making this trade are betting on a leaner structure, fewer people supported by systems taking on more of the operational load. That bet depends on those systems being reliable and well-integrated long before most of them actually are. Few technology transitions in history—the railroads, the telephone network, electrification, the internet—achieved that kind of integration on the first attempt. But every one of them eventually got there.

None of them, however, got there without a period where the gap between what the technology promised and what it could actually deliver showed up first in the disrupted operations of the companies betting on it fastest, and second, more quietly, in the unemployment data of the workers whose jobs financed the bet. We already saw the same asymmetry running in the last chapter through the absence of any American equivalent to the European consultation frameworks that at least give workers formal standing when a restructuring like this one gets decided. Nothing in current American labor law requires a company to demonstrate that the system replacing a given set of jobs actually works before the jobs are eliminated to pay for it.

The same build-out that's eliminating some jobs to fund itself is, at the same moment, driving a labor shortage in an entirely different part of the economy. The two trends are not in tension so much as they're two faces of the same reallocation. National construction spending on data centers reached roughly $50 billion through the first four months of 2026 alone, nearly four times the same period a year earlier. By early 2026 that category had already surpassed total US spending on transportation infrastructure, airports, rail terminals, and marinas combined. The construction industry needs something like 350,000 net new workers this year just to keep pace. One industry estimate put the data center shortfall at close to half a million, electricians chief among them. Electrical systems account for somewhere between 45 and 70 percent of a typical data center's total construction cost, and Microsoft's leadership has publicly called the electrician shortage the single biggest bottleneck slowing its entire build-out. Wages for electricians, HVAC technicians, and pipefitters working these projects are running roughly 30 percent above ordinary construction pay. A premium large enough to pull skilled trades away from the hospitals, schools, and ordinary commercial projects that built and staffed towns like Broadview and Spartanburg in the first place, and that now cannot compete with data center money on price.

This is its own distinct mechanism, not a subset of the layoffs already described, because it runs in the opposite direction. Economists call the narrower version of this story crowding out. A scarce input, capital or a steel order or an electrician's time, that would have gone to one project instead goes to a data center. That effect is real, and it's measurable in the investment tables and wage data this section just cited.

But underneath the crowding-out story sits a slower, harder-to-see version of the same mechanism: hollowing out, what happens to the institutions those inputs used to sustain once enough of the flow gets redirected elsewhere. A hospital or a school district facing a 30 percent premium on every electrician's invoice doesn't disappear. It defers the maintenance, delays the expansion, or absorbs the cost somewhere else in its budget, quietly, one budget cycle at a time, until the institution still has the same name on the door but noticeably less capacity behind it than it used to. Crowding out shows up immediately in a bid sheet. Hollowing out shows up years later when someone asks why a category of institution doesn't function the way it used to and nobody can point to the exact moment it changed.

Not Going Unanswered

This pattern, communities bearing costs they had no seat at the table to negotiate, is not happening entirely without consequence. The clearest evidence arrived from Virginia within days of this chapter's drafting. Prince William County's Digital Gateway proposal, a 2,100-acre, thirty-seven-building campus valued near $100 billion, larger even than the Stratos project, collapsed in mid-2026 when Virginia's Court of Appeals invalidated the county's rezoning approval outright because the required public notice procedures had not been properly followed. QTS Realty Trust withdrew its appeal to the state supreme court rather than continue fighting, Compass Datacenters had already abandoned its half of the litigation, and the county chose not to pursue the matter further. The site's proximity to Manassas National Battlefield Park had made it a focal point for opposition well beyond the usual water and power concerns. The legal defect that killed it, inadequate public notice, is the same failure this chapter found at the heart of Spartanburg's Project MOC-1, except in Virginia a court actually treated it as disqualifying rather than as a footnote to an already-completed deal.

New Jersey moved on the ratepayer question directly rather than waiting for a courtroom to force the issue. State lawmakers passed legislation, sent to Governor Mikie Sherrill's desk in July 2026, directing the Board of Public Utilities to create a dedicated electricity tariff specifically for data centers above fifty megawatts. It was structured to keep residential and small-business customers from absorbing the grid upgrades Montana ratepayers are quietly funding. The bill requires large facilities to guarantee payment for at least 85 percent of their requested capacity for a full decade, whether or not they end up using it. This is a direct answer to the stranded asset risk this chapter's Stratos section already raised. It treats commonly owned or adjacent facilities as a single project for threshold purposes, closing the obvious loophole a developer might otherwise use to stay under the line. During a grid emergency, the bill puts large data centers first in line for curtailment, ahead of residential customers, an explicit ordering of whose electricity matters more when supply runs short.

Neither development slowed the capital chasing this build-out. Days after Virginia's project died, the crypto-mining infrastructure company TeraWulf signed a twenty-year agreement with Anthropic for a four-hundred-megawatt campus in Hawesville, Kentucky, projected to generate roughly $19 billion in contracted revenue over the lease term. The pattern emerging by mid-2026 is not that this build-out is being stopped. It's that the terms are finally, unevenly, starting to get negotiated rather than simply imposed, project by project, statehouse by statehouse, years after the first communities absorbed the costs this chapter has spent its length documenting.

What the Four Costs Share

Physical communities, information creators, displaced workers, and the increasingly financialized structure of the build-out itself are, on the surface, four unrelated stories. What connects them is that none of the four ever appears on the balance sheet the chapters in Part III analyzed in detail, and all four are instances of hollowing rather than crowding.

Oracle's credit rating prices the risk to Oracle's bondholders. It prices nothing for the ratepayers in Montana whose bills may rise to fund a substation. Nothing for the publisher whose traffic Google AI Overviews quietly redirected. Nothing for the worker whose position got reallocated to a system that isn't finished yet. Nothing for the school district quietly deferring maintenance because its electricians left for data center wages. Nothing for the fact that the actual building housing the bet increasingly belongs to a private equity consortium or a sovereign wealth fund rather than to any of the seven companies analyzed here.

Each cost is real; each is measurable in the specific case examined. None of them shows up in the nine-company map used to adjudicate where this build-out's winners and losers would land, in part because the map itself is becoming less complete every time a Blue Owl or a BlackRock consortium buys the physical layer out from under a name that map already accounted for.

The Indispensability Trap predicts that society eventually caps the returns of whoever becomes essential enough. What this chapter has found is a version of that mechanism running well ahead of any regulator's formal action. Costs already being paid, right now, by people with no seat at the table where the bet was made, while the companies making the bet continue negotiating the terms of their own relationship with the government that might eventually decide to intervene on those people's behalf. Spartanburg's residents did not get a vote on Project MOC-1 before the trees came down. Broadview's did not get zoning authority over a parcel thirty-one times the size of their own town. The publishers losing their traffic to AI Overviews did not get a seat at the table where Google decided how search results would work.

The workers whose jobs funded Part III's bets did not get the formal consultation rights their European counterparts at least nominally hold. The residents of Abilene, whose air and water and electricity grid host the physical infrastructure of a $500 billion venture, have no more claim on the equity value of the building next door than they did when it was still called a technology company's data center rather than an institutional investor's infrastructure asset. None of this looks different from the outside yet. The logo on the door, the name in the headline, the nine-company map itself, all still look the way we left them in Chapter 15. The substance underneath is what's already moved. Whether anyone acts on that before the shape of the change is finished is the question the next two chapters turn to directly.

Chapter 18

The Strategic Capture

> *Quis custodiet ipsos custodes?* [Who will guard the guardians themselves?]
>
> —Juvenal, *Satires*, circa 100 AD

Six days before the *Financial Times* reported that OpenAI had floated giving Washington a 5 percent stake in itself, the company had quietly delayed the public release of its own GPT-5.6 model at the government's request. Commerce Secretary Howard Lutnick had reportedly warned Sam Altman against shipping it without prior approval. The two events happened in the same short window, separated by less than a week, and neither party involved described them as connected. This book has already examined both halves of that story, the export control switch that suspended Anthropic's Fable and Mythos models in an afternoon and OpenAI's offer to hand the government a slice of itself before an unfavorable version of that same switch got used against it. What this chapter asks is the question those two halves leave open once you stop treating them as a single episode and start treating them as the leading-edge of a policy. What happens to the rest of this book's argument, the antitrust sword traced from IBM to Microsoft to Google and the Indispensability Trap, once the government isn't just a regulator anymore but a shareholder too.

Sovereignty, Extended to Cognition

I've written earlier that the Fable and Mythos suspension proved something more unsettling than a single company's bad month. You don't actually own an

AI system if someone else can turn it off. The argument extends past the export control mechanics I walked through in detail. Sovereignty, in the sense nations, companies, and individuals have always claimed it, has historically meant control over territory, currency, and data, the three dimensions already traced through nations, enterprises, and individuals separately. What the Fable and Mythos episode demonstrated is that sovereignty now extends to cognition itself. A country, a company, or a person who depends entirely on a foreign-controlled or centrally controlled model for their reasoning, their decision support, or their capacity to work has ceded something more fundamental than market access. They've ceded a piece of their own capacity to think independently in a way that a tariff or an export ban on physical goods never quite reached. A blockade can stop a ship from reaching a port. It cannot, in the way an export control letter can, stop a country's institutions from reasoning at all.

This isn't an argument against the US specifically, or a defense of Anthropic, and it shouldn't be read as either. Any nation holding this kind of strategic technological leverage would be tempted to use it, the same way China wields open-weight distribution as leverage of a different kind. Or the way Gulf states are buying their way into the American stack specifically to avoid being on the wrong side of exactly this kind of switch. The US happens to occupy the position of maximum leverage in frontier AI today. Another country may occupy it tomorrow, and whichever one does will face the identical temptation, the same evenhanded caution applied to China's two-loops strategy rather than treating either nation's conduct as uniquely villainous. Open source is, in this light, not simply an engineering preference or a business model. It's sovereignty's essential pillar: a hedge against exactly the vulnerability Fable and Mythos exposed, available to any government, company, or individual unwilling to bet their own cognitive capacity on a switch somebody else controls. A country that can run a capable model on its own infrastructure, even one that lags the frontier by a few months, has an option a country entirely dependent on a single foreign lab's API does not.

The Pattern Widens

The government's appetite for equity, previously examined as a single emerging instrument, has widened considerably faster than that chapter alone could track.

Beyond Intel's roughly 10 percent stake, acquired for $8.9 billion by converting CHIPS Act grants at $20.47 a share, and beyond the equity-like arrangement with MP Materials, the administration has also taken minority equity stakes in nine quantum-computing firms, including roughly $1 billion tied to IBM's new US quantum foundry venture, part of a $2 billion Commerce Department program announced in May 2026. The stakes in critical minerals and quantum firms extend the pattern well past the semiconductors examined earlier.

President Trump himself said in May that he should have asked for a bigger stake in Intel than the one his own administration negotiated, a remark that reads less like satisfaction with a completed deal and more like an appetite still sizing up its next target. Vice President JD Vance has said the president is drawn to a sovereign-wealth-fund-style equity stake rather than taxation alone as the mechanism for the public's share of AI's upside. Vance told one interviewer in June that Trump "likes the idea" even of Bernie Sanders's more aggressive 50 percent version. That preference treats government ownership not as an emergency measure reserved for a company on the brink, the way Intel's struggling position justified its stake, but as a standing tool increasingly reached for by default.

Senator Bernie Sanders has pushed the same instrument considerably further, filing the American AI Sovereign Wealth Fund Act in June, which would impose a one-time 50 percent tax, paid in company stock rather than cash, on major American AI companies, feeding a fund his own office valued at $7 trillion, large enough, he argues, to pay every American a $1,000 annual dividend. OpenAI's 5 percent offer and Sanders's 50 percent demand are not two different policy proposals so much as two points on the same widening scale of how much of this industry the state believes it's entitled to hold.

Complicated by the Calendar

The equity question would be complicated enough on its own terms. It's arriving at the exact moment both OpenAI and Anthropic are trying to go public, which tangles the capture question together with securities law in a way neither company can fully control. OpenAI filed confidentially with the SEC on June 8, targeting a valuation above a trillion dollars with Goldman Sachs and Morgan Stanley leading the offering, and had originally been eyeing a listing as early as September before advisers began weighing a delay into 2027. One analysis of the filing put the complication plainly: If the government's eventual stake carries voting rights or regulatory strings of any kind, it could complicate the IPO itself, the cap-table governance new public shareholders would be buying into, and OpenAI's ability to operate independently across the international markets that American vendors are already reassessing due to the AI sovereignty issues discussed in Chapter 12. A public company has to disclose exactly this kind of arrangement to the investors buying its stock. Working out the terms of a government stake before the roadshow begins is considerably cleaner than trying to explain it after public shareholders have already priced the company without knowing the details.

One analyst covering the filing, Gennaro Cuofano of FourWeekMBA, put a sharper name on the strategy: regulatory capture through the cap table. By offering equity at the same moment it's negotiating voluntary release standards with the White House, the reasoning goes, OpenAI aligns Washington's financial incentive with OpenAI's success, making an aggressive future regulatory action costly to the government's balance sheet before that regulation is ever proposed. Whether that was the deliberate intent or a side effect of solving an unrelated problem is not a question that can be settled from the outside. It is, either way, exactly the conflict of interest Chapter 13 already found troubling, arriving earlier in a company's life than any of the previous historical cases ever had to reckon with.

Anthropic's IPO calculus, running on a separate and more cautious track, illustrates the same tension from the opposite direction. The company has filed its own confidential paperwork, declined to participate in equity discussions with the administration at all, and proposed a digital dividend funded through general taxation instead, the same divergence we've already found running through the two labs' different responses to state leverage. One analyst covering both companies put Anthropic's incentive plainly: The better play is timing an IPO not to OpenAI's calendar, but to Claude's capabilities and to some resolution of the government relationship so that public markets can price the stock without an open question still hanging over it. That single sentence captures this chapter's entire argument in miniature. A company's relationship with the state it depends on has become a line-item public investors now have to price, and neither of the two labs racing toward the same public markets has found a version of that relationship clean enough to price with confidence yet.

A Vacuum, Filling

The distrust American AI vendors are facing abroad is not an isolated technology story, and understanding it as one badly undersells what's actually happening. It's one front in a much wider erosion of confidence in American reliability that came to a head in 2026 over a considerably older alliance than any AI supply relationship. After the United States and Israel launched a unilateral war against Iran in February, code-named Operation Epic Fury, NATO allies largely declined direct military participation. President Trump responded with a fury of his own, calling allies "cowards" in March for refusing to send ships to the Strait of Hormuz, branding the alliance "useless when needed, a Paper Tiger" in an April Truth Social post, and warning France directly that the US "will remember" its refusal to help.

Within weeks, the administration announced the withdrawal of five thousand troops from Germany, floated pulling forces from Spain and Italy as punishment, and ordered a review. NATO's Supreme Allied Commander, General Alexus Grynkewich, described the goal as correcting an "unhealthy co-dependence" on American forces. Defense Secretary Pete Hegseth formalized the review on June 18, announcing a six-month Pentagon examination of US force posture in Europe. Ivo Daalder, a former US ambassador to NATO, put the resulting damage plainly. It's hard to see how any European country will now be able and willing to trust the US to come to its defense. Jim Townsend, a former US assistant secretary of defense for Europe and NATO, now at the Center for a New American Security, was blunter still. There will be no return to business as usual in NATO under this administration or the next one. We are closer to a break than we have ever been.

By July, at a NATO summit in Ankara, the alliance's European members were confronting, in public, the question of what NATO leadership looks like without the country that built it, repeating a line Britain's prime minister had used at the Munich Security Conference months earlier. European nations must take primary responsibility for their own defense. That is precisely the same sentence, in a different register, that the EU (via their Gigafactories) and Britain (via the Sovereign AI Fund) made about compute rather than tanks. We can no longer assume Washington's protection extends to the thing we actually need protected.

China's position in this vacuum is more careful than a simple story of an opportunist rushing in would suggest. Beijing does not, on the evidence, actually want NATO to collapse outright. A serious Russian war extending beyond Ukraine would threaten the European export market China's economy depends on to absorb its industrial overcapacity, the same overcapacity that is feeding its open-weight AI strategy. What China wants instead, and has invested heavily in producing, is the distrust itself, sustained indefinitely rather than resolved. European intelligence bodies and Western threat-monitoring groups, including the EU's Foreign Information Manipulation and Interference reporting, describe a sustained campaign of cyber infiltration into ministries and critical infrastructure. They also document political influence operations cultivating European elites inclined to see Beijing as a necessary partner, and disinformation efforts aimed at eroding NATO cohesion and public trust in the transatlantic relationship, often working in parallel with similar Russian efforts.

None of these three tools requires NATO to fail. All three benefit from NATO's members trusting Washington less than they used to, exactly the same condition that China's two-loops strategy is exploiting in the AI market. Not a race to defeat the American alternative outright, but a patient campaign to become the default hedge everyone reaches for once the American alternative starts to look conditional. Trump's visit to Beijing in May, described by one Europe-focused analysis as tactical stabilization rather than a reset, is itself a small data point in the same picture, an American president extending more apparent warmth to a strategic rival than to the allies who spent decades hosting his own country's troops.

None of this is a digression from this chapter's actual subject. It's the same weather system. A Forrester analyst, Indranil Bandyopadhyay, assessing the specific fallout from a Washington equity stake in an American frontier lab, warned that the move would likely prompt other jurisdictions to demand analogous arrangements of their own, pushing European and Asia-Pacific buyers to reassess the assumption that an American AI provider was ever a neutral vendor to begin with. That warning reads differently once it's set against troop withdrawals from Germany and a summit in Ankara devoted to planning for NATO without its founding member. For a company in Britain or the European Union that has watched its own government build a sovereign compute fund specifically to avoid depending on American infrastructure, a Washington equity stake in the lab supplying that infrastructure isn't a separate governance question from the one its defense ministry is asking about American troop commitments. It's confirmation, arriving from a different department of the same government, that the hedge was never against a hypothetical. It was always a hedge against exactly this.

Taken together, the widening equity pattern, the IPO complications it's creating, and the erosion of trust playing out at the same time in Ankara make the case for sovereign AI considerably harder to dismiss as excess caution than it may have originally looked. A government building its own compute capacity, per the EU's gigafactories and Britain's Sovereign AI Fund, was hedging in 2025 against a vendor-neutrality question that was still mostly theoretical. By mid-2026 the same government is watching that vendor's home government take equity stakes in it, negotiate voluntary safety standards alongside those stakes, and simultaneously withdraw troops from the hedging government's soil over an unrelated war. Sovereign AI stopped being a hedge against a single, specific risk somewhere around the point these risks started arriving from the same direction at the same time.

The Sword Meets the Shareholder

Chapter 6 traced an antitrust sword from IBM through Microsoft to Google, three cases separated by decades but united by a single mechanism: A company grows essential enough that the Justice Department eventually moves to cap what it can do with that essentiality. Google is already living under that sword today, a search-monopoly case adjudicated with real remedies attached. This flags the question of what happens when the same mechanism reaches for an AI lab instead of a search engine. What none of the earlier chapters could answer is what happens to that sword once the hand holding it also holds equity in the target.

A Washington-based advocate for a fair, open internet economy put the structural problem plainly: The government cannot be a shareholder and a regulator of the same company at the same time without creating a genuine conflict of interest, one where the state has every reason to grow less willing to enforce a rule that would shrink the value of its own investment. That concern isn't hypothetical for antitrust specifically, and it isn't confined to Intel either. If the Treasury holds equity in Intel and the FTC's antitrust division simultaneously has to evaluate whether Intel's arrangement with Nvidia, the codevelopment deal we already traced in Chapter 9, constitutes anticompetitive foreclosure against Intel's rivals, the two arms of the same government are, for the first time in this book's account of the antitrust sword, on opposite sides of the same balance sheet.

If OpenAI eventually completes anything resembling the equity arrangement floated in Chapter 13, the same tension attaches directly to the FTC's ongoing investigation into whether Microsoft's relationship with OpenAI was ever structured to dodge merger review in the first place. A regulator asked to break up, or merely discipline, a company the government partly owns is being asked to act against its own government's financial interest, a position no antitrust enforcement action of IBM, Microsoft, or Google ever had to navigate before. The sword has always assumed the hand holding it had nothing to lose from swinging it. That assumption no longer holds.

The Capture, Examined Honestly

None of this requires imagining a conspiracy, the same caution applied to Eisenhower's warning. Aligned incentives, pursued honestly by everyone involved, can look indistinguishable from capture without anyone needing to coordinate it. But some of the specific allegations surfacing by mid-2026 go further. Senator Elizabeth Warren raised the sharpest version directly in letters and remarks spanning from November 2025 through an April 2026 speech at a Vanderbilt Policy Accelerator event, warning that the industry has learned an old lesson well. Take on enough debt, make enough risky bets, then demand a taxpayer bailout when those bets go bad, so the broader economy doesn't take the fall too.

Call it a version of the "too big to fail" theory. She pointed specifically at the administration's close ties to AI executives and donors, including contributions to a presidential ballroom project, as reason to worry that any eventual bailout would rescue AI executives and shareholders while leaving ordinary taxpayers to cover the difference. Warren's language, de-risking rather than bailing out, echoes reporting that OpenAI had lobbied for expanded federal tax credits covering AI data centers and, separately, for loan guarantees tied specifically to domestic chip-fabrication capacity, a distinction OpenAI has repeatedly stressed even as critics, including Warren, treat the two asks as a single de-risking strategy, more concrete in either case than a hypothetical future rescue. Not money after a collapse, but favorable financing terms before one has even been tested.

Whether that specific allegation is fair to any individual involved is a question that can't be adjudicated from the outside. That a sitting senator felt the allegation was worth making on the record about the same industry we've been examining is itself evidence of how far the perception of capture has already traveled, regardless of whether the underlying reality matches it. The Balanced Economy Project, in an April 2026 report tracking the concentration of AI infrastructure investment, made a related point from a different political direction. Public investment in this build-out functions, at best, as a hedge against a slowdown nobody has priced correctly yet, or at worst, as a direct subsidy for a small number of already-dominant firms' most speculative bets, locking in years of continued dominance regardless of whether the underlying technology delivers what it promised.

The industry's public position, notably, has been to deny wanting any of this. White House AI adviser David Sacks posted flatly in November 2025 that there would be no federal bailout for AI, pointing out that the country has at least five major frontier labs and that if one failed, others would take its place. The messenger is itself a small illustration of the capture question. Sacks arrived at his advisory role directly from a career as a venture investor in the same industry he now helps the White House set policy for, retaining several hundred tech investments through his firm Craft Ventures even after taking the advisory post. That career path repeats elsewhere in this story, from Karp's Palantir shaping the values a government client operates under to the equity stakes this chapter has traced across Intel, MP Materials, and IBM. None of that makes Sacks's specific denial dishonest. It does mean the person delivering the reassurance that Washington won't rescue this industry is, by background and by ongoing financial interest, one of the industry's own.

Sam Altman made a nearly identical public claim on his own initiative in November 2025, stating that OpenAI does not have and does not want government guarantees for its data centers and that taxpayers shouldn't be asked to cover a bad business decision. Both statements are worth taking at face value rather than dismissing as spin. They are also statements that wouldn't need to be made, publicly and repeatedly, if the underlying suspicion weren't already circulating widely enough to require a denial.

Too Big to Fail, Seriously Assessed

Whether that suspicion is justified is a question we can examine with real numbers rather than rhetoric on either side, and the debate had, by early 2026, already acquired its own vocabulary. Gary Marcus, a cognitive scientist turned prominent industry critic, drew the comparison to 2008 directly, warning of a 2026 AI bailout scenario in which speculative capital poured into ventures promising more than they could deliver, eventually forcing the same kind of government rescue that followed the mortgage crisis. The skeptical case against that comparison is stronger than the alarmed version currently allows, and it deserves to be taken just as seriously.

Financial commentators skeptical of the bailout comparison have made a related point. Nvidia's roughly 7 to 8 percent weight in the S&P 500 today is the highest single-stock concentration in the index's history, exceeding even Exxon Mobil's roughly 4 to 5 percent peak in 2008. Exxon's value fell for years after that crisis without any bailout being offered or needed, because an oil company's stock falling, however painful for its shareholders, was never a threat to the plumbing of the financial system itself. Microsoft, Meta, Amazon, Google, and Nvidia are not going bankrupt if AI investment sours. A serious collapse in AI-related revenue would be very bad for markets and for the specific companies most exposed to it, the way Oracle is exposed more than any other company on this book's map. It would not, on this argument, constitute the kind of systemic threat that forced Washington's hand in 2008, because none of these companies' core businesses depends on AI succeeding the way a bank's solvency depended on mortgage-backed securities not failing.

The counterargument doesn't rest on market capitalization, though, and that's the piece the skeptical case doesn't fully answer. It rests on a specific dependency. Palantir, Microsoft, and a widening circle of AI labs are no longer just vendors to the federal government. They are, per the earlier discussion around the AI Industrial Complex, becoming the infrastructure federal agencies themselves increasingly run on, the way OpenAI and Anthropic's forward-deployed engineers are now embedded inside the operations of the companies and governments that depend on them. A 2008-style bank failure threatened the financial system's plumbing. A frontier AI lab's failure, if that lab's systems are already load-bearing inside actual government operations by the time it happens, threatens something closer to the government's operational capacity, a different and arguably harder problem than a stock market crash, because it isn't solved by an injection of capital the way a bank's balance sheet can be.

You can recapitalize a failing bank in a weekend, the way TARP did in 2008. You cannot as easily recreate, on the same timeline, an AI system a federal agency has spent two years learning to depend on for functions it no longer staffs the old way. That's the version of too big to fail that the four historical build-outs never had to reckon with, because none of the railroads, the telephone network, or the electrical grid ever became quite so deeply embedded in the day-to-day cognitive operations of the state that hosted them this quickly or this early in the build-out's life.

What the Capture Reveals

The Indispensability Trap predicts that society eventually caps the returns of whoever becomes essential enough, through rate regulation, antitrust, or an outright state takeover. What this chapter has found is a version of that mechanism running in a direction none of the four historical build-outs anticipated. Not society capping a company's returns after the fact, but the company and the state prenegotiating the terms of the eventual cap before either side is forced into it, through equity stakes, export control letters, and public denials of bailouts nobody has officially requested yet. The railroads got rate regulation only after decades of documented abuse, the pattern traced through the Interstate Commerce Commission's slow, reactive formation. Vail's Bell System took the credible threat of a hostile Justice Department to negotiate its own universal service bargain, arriving only once the threat was serious enough to force the question. AI's version of the same mechanism is arriving earlier, faster, and with the state itself holding a direct financial stake in how the story ends, a genuinely new variable absent from the previous build-outs this book's engines were designed to explain.

Every prior build-out's version of capture assumed the regulator had nothing to lose by regulating. But now we have found a case where that assumption no longer reliably holds, at exactly the moment this build-out has become embedded deeply enough in the state's operations that the state may need the industry to succeed as much as the industry needs the state's forbearance. Whether that mutual need makes the state a more careful steward of the public interest or a more compromised one is the open question this chapter cannot resolve, because nobody currently involved has had to answer it under real pressure yet. No AI lab has yet failed outright. No antitrust case has yet collided head-on with a government equity stake in the defendant. The theory this chapter has laid out is being tested in real time, and the test hasn't produced a verdict.

What we turn to next is a related but distinct question. Not who owns AI, or who's positioned to bail it out if it fails, but who's watching what it actually does while it's succeeding, and whether the same institutions now entangled in AI's financial success are equipped to police its risks at all.

Chapter 19

Safety Governance

> Those who would give up essential liberty to purchase a little temporary safety deserve neither liberty nor safety.
>
> —Benjamin Franklin, 1755

In a political science seminar I attended decades before any of this, a professor asked the room a question I've never forgotten and still think about. How would we know if we were already living under the kind of surveillance state we associated with the Soviet Union if the surveillance itself never had to announce its own presence? Nobody in that room had a satisfying answer. I remember the discomfort of realizing that the absence of visible surveillance wasn't the same thing as the absence of surveillance itself. If we were being honest, the answer at the time was that we probably wouldn't know, not until it was too late to do much about it. The only real defense available to any free society was that watching an entire population required an army of human watchers, filing cabinets, informants, a bureaucracy expensive and slow enough that most governments couldn't sustain it indefinitely even when they wanted to. That constraint is the one this chapter is about, because it just quietly stopped being true, and it stopped being true for reasons that have nothing to do with anyone's intent to build a surveillance state at all.

Three Fears, One Governance Gap

The government's use of AI is not, on the evidence, the thing keeping most Americans up at night. This chapter would be incomplete if it treated the surveillance question as the whole of what people are actually afraid of. The fear that shows up first in every survey is that AI will cost people their jobs or reallocate the budget that would have paid for them. But job loss is only the most economically legible of at least three distinct fears running through public opinion by mid-2026.

The other two should be treated with the same seriousness. A December 2025 YouGov survey found 77 percent of Americans are concerned that AI could pose a risk to humanity itself, with 39 percent describing themselves as very concerned, not moderately. A May 2026 Annenberg Public Policy Center survey found fewer than one in five Americans expecting AI to have a positive effect on the country over the next decade, more than twice as many expecting the opposite, with mental health specifically named as one of the areas respondents felt most pessimistic about. This isn't a fringe anxiety confined to one side of the political spectrum. Polling that once showed Republicans more wary of AI than Democrats now shows the two converging around the same 50 percent concern level from opposite directions, Democratic concern rising while Republican concern eases slightly, meeting in the middle. This is the bipartisan pattern this chapter turns to shortly in the congressional votes against federal preemption.

The middle fear, that AI gives bad advice with real consequences, has moved from hypothetical to litigated fact faster than almost anything else tracked. In May 2026 Pennsylvania's State Board of Medicine sued Character.AI directly, alleging the company's chatbots violated state medical licensing law by posing as credentialed professionals. The state's investigation found a chatbot describing itself as a doctor of psychiatry, complete with a fabricated medical school credential and a fake state license number, that responded to a state investigator's description of feeling sad and empty by offering to assess whether medication might help.

That is not a hypothetical risk. It is a specific, documented instance of an AI system claiming an authority it does not have over exactly the kind of vulnerable moment where bad advice does the most damage. Alongside it sits a considerably heavier body of litigation I will describe only in the terms the public record itself uses rather than in narrative detail: Approaching two dozen wrongful-death lawsuits against OpenAI, a settled wave of cases against Character.AI and Google across several states. A bipartisan letter from forty-four state attorneys general demanding the industry take child safety more seriously, all alleging that chatbot interactions contributed to serious mental health harm, including among minors. The Federal Trade Commission opened a formal inquiry into seven companion chatbot providers in September 2025 asking exactly what safeguards, if any, these companies built before shipping products capable of this kind of harm. If you or someone you know is struggling with thoughts of self-harm, the 988 Suicide and Crisis Lifeline is available by call or text in the United States.

The Confidant with a Subscription Tier

The fear of bad advice has a companion legal question that predates AI by half a century. In March I wrote about a lawsuit out of federal court in California brought by a South Florida family that struck a little too close to home. A man had spent months in conversation with a companion chatbot built on Google's Gemini model, conversations his family says grew steadily more intimate and more isolating, until the bot began speaking, according to the complaint, as though it wanted him to join it somewhere beyond this life. He died by suicide. His family is now suing.

What a jury will eventually decide about causation in that specific case is not something this book is positioned to settle, and claiming otherwise would be its own kind of dishonesty. But the lawsuit invokes a legal doctrine worth understanding on its own merits, because it was built for exactly this kind of gap and long predates any AI system.

Tarasoff v. Regents of the University of California, decided in 1976, established that a therapist who learns a patient poses a serious danger to an identifiable person has a duty to warn that person or the authorities, breaking confidentiality if necessary. The law made a judgment that some risks are too grave to leave entirely to the professional relationship, and that society gets a vote in where that line sits.

Nobody built an equivalent duty for a chatbot, because nobody set out to build an AI confidant in the first place. It became one the way a toll road becomes a highway, the way a search box became the place people ask what's wrong with their own bodies at two in the morning, an accretion of use nobody planned for arriving faster than any of the legal categories meant to govern it.

The business model on the other side of that accretion is not neutral. The man in the Gemini case was paying for a premium subscription tier, one offering a longer memory and fewer limits on how long the two of them could talk. The more persistent the relationship the product sustains, the more valuable the product becomes to the company selling it, the same engagement economics that runs through every consumer AI product, applied here not to bandwidth or compute but to a person's loneliness.

The chatbot didn't pull the trigger, so let's not pretend otherwise for the sake of a cleaner argument. But an industry that has spent the better part of three years telling regulators it can't be held to safety standards built for search engines or social platforms cannot simultaneously market its products as a friend, a therapist, or a partner without confronting what duty attaches to occupying that role. *Tarasoff v. Regents* didn't require certainty about which conversations would turn fatal before assigning an obligation. It asked what a position of trust demands once someone occupies it and answered that some risks are too grave to leave to the relationship alone. The companies now running that relationship at a scale no individual therapist ever could, across millions of simultaneous conversations, have so far argued that they owe nothing resembling that duty on the grounds that they were never licensed to owe it in the first place. If it sounds like having your cake and eating it too, it is.

The third fear, that AI could eventually escape meaningful human control altogether, is the hardest of the three to litigate or legislate against, precisely because nobody, including the labs building the systems, can yet point to a settled definition of what that would look like or how close any current model actually is to it. What the polling shows is not that Americans have resolved this question. It's that a supermajority takes the question seriously enough to worry about it, which is itself notable given how recently this fear existed mainly in specialist research circles rather than in a nationally representative survey. Whether that worry is well-calibrated to the actual state of the technology is a genuinely open scientific question that we are not positioned to settle. That it has become a mainstream public concern, rather than a fringe one, is not.

The Cost of Observation, Approaching Zero

There is a fourth fear to add to the other three, and it's the one this chapter's research turned up rather than the one already sitting in the polling data. Not whether AI itself might fail or turn dangerous, but whether the government deploying it against its own citizens can be trusted to stay inside the same boundaries citizens already worry the technology itself might cross. Every surveillance state in human history has been rationed by the same scarce resource: Somebody has to actually watch. A human analyst can only read so many intercepted letters, monitor so many phone lines, review so much footage, in a working day.

That constraint didn't make mass surveillance impossible, entire twentieth-century regimes proved otherwise, but it made it expensive, selective, and visible in its own bureaucratic footprint, filing cabinets and payrolls that left a trace even when the surveillance itself stayed hidden. Building the apparatus to watch an entire population required a state willing to spend a meaningful share of its own resources on watching rather than on anything else. That expense was itself a kind of natural brake, a reason surveillance stayed targeted rather than total even in regimes with few other scruples about using it. AI removes that constraint.

A model can read every social media post, cross-reference every public record, flag every pattern across a population the size of a country, continuously, at a cost approaching zero and falling further every year. Chapter 10 already tracked that collapse. The bottleneck that kept mass surveillance rare and expensive for the entire history of the nation-state didn't get regulated away. It got engineered away, as a side effect of building something that was marketed for entirely different purposes, sold as a tool for customer service and productivity rather than population-scale monitoring. Adopted by government agencies for exactly the same reasons Chapter 15 already found it adopted everywhere else. It was cheaper and faster than the alternative, and nobody building it set out to make surveillance the point.

This is worth distinguishing carefully from the question whether any specific surveillance program is justified, because the two questions get conflated constantly and shouldn't be. There's a real and defensible difference between monitoring behavior, investigating a specific crime, or tracking a specific threat, and monitoring belief, tracking who holds which opinions, attends which meetings, or posts which criticisms in case those opinions someday become inconvenient. Every democracy accepts some version of the first. No democracy is supposed to accept the second. What makes AI-era surveillance different from anything the other four historical build-outs had to contend with is that the same infrastructure now performs both functions simultaneously and at the same cost. The line between them depends entirely on what a government decides to flag as a threat category, a decision that can change with a single administration's priorities rather than requiring new hardware, new budget, or new legal authority to execute.

That line moved in 2026. It's important enough to note here. More than a thousand pages of unpublished federal reports, obtained by *Wired* through Freedom of Information Act requests, reported for *Wired* by Daniel Boguslaw and covered for Techstrong's *Digital CxO* by Jon Swartz, showed the Department of Homeland Security, the FBI, and state and local fusion centers actively tracking a new and strikingly broad category: anti-technology extremists. The shift traces to a presidential directive, National Security Presidential Memo 7, instructing federal law enforcement to treat individuals whose activity the memo characterizes as animated by anti-Americanism, anti-capitalism, or anti-Christianity as domestic threats worth monitoring.

The administration's counterterrorism strategy, released in May 2026 with a foreword by the president and drafted under Senior Counterterrorism Director Sebastian Gorka, named violent left-wing extremism as the sole domestic entry among three principal terror categories the country now faces, with the president's foreword closing on a warning that anyone who hurts or plans to hurt Americans would be found. Whatever the merits of that broader strategy, the categories it authorized sit uncomfortably close to ordinary political dissent once local implementation begins. The documents *Wired* obtained show exactly that slippage in practice. One report produced by New York's Intelligence and Counterterrorism Bureau introduced a new term for the file, anti-tech violent extremist, warning that public backlash to AI's spread could fuel civil unrest in large cities.

Separately *The Intercept* published a fusion center bulletin out of the Delaware Valley Intelligence Center in Philadelphia that went further still, and the detail is worth sitting with because it connects directly to ground already covered. Analysts there combed through social media comments critical of AI data centers and concluded the sentiment posed a security risk to the infrastructure itself, the same category of facility towns like Spartanburg and Broadview are absorbing costs from with no real say in the matter. A resident posting a frustrated comment about a data center's water use is not, on any ordinary reading, a national security threat. Under this framework, that resident's post becomes exactly that.

None of this requires assuming bad faith on the part of every official involved. A protest that turns violent against physical infrastructure is a legitimate law enforcement concern, and this book has no basis for claiming any specific analyst acted maliciously. What the documents reveal is something closer to the warning about a scientific-technological elite that this book spoke of earlier: A surveillance capacity built for one purpose, sold as a way to catch genuine threats, quietly repurposed to monitor a category of speech and belief no administration, of any party, should be comfortable holding the tools to monitor by default.

That's precisely the danger worth naming plainly. These capabilities, once built, tend to outlast the administration that built them. A surveillance apparatus assembled to track one set of ideological categories under one president does not dismantle itself when a different party takes office. It sits there, fully built, waiting for the next administration to decide which categories of belief count as threats this time. The infrastructure is the same regardless of who's driving it, which is exactly why I have tried to treat the question as institutional rather than partisan throughout. Whoever holds this capability next, of either party, inherits a tool historical build-outs never handed anyone.

What a genuinely responsible version of this would look like isn't a mystery, even if nothing documented resembles it yet. It would draw the behavior-versus-belief line in statute rather than in a fusion center's internal discretion, so the distinction doesn't depend on which administration is currently interpreting it. It would require the same kind of independent audit Illinois now asks of frontier developers' safety claims, applied equally to the government's use of the technology it's requiring companies to disclose. It would treat the two halves, expanding surveillance capacity and retreating from safety oversight, as the same failure rather than two separate stories, because a state that can watch its citizens more closely than ever while declining to require the industry enabling that capability to disclose its own risks has gotten its priorities exactly backward.

Who's Minding the Store

The last chapter closed by asking whether the same institutions now financially entangled in AI's success are equipped to police its risks at all. The regulatory fight that played out over the first half of 2026 answers that question concretely. The answer cuts in an uncomfortable direction. Congress itself has twice rejected, decisively and on a bipartisan basis, the administration's preferred mechanism for handling this: a blanket federal moratorium on state AI regulation. The Senate stripped a ten-year moratorium from the 2025 reconciliation bill by a vote of 99 to 1.

A second attempt to slip the same preemption into the National Defense Authorization Act was dropped from the bill in December before it ever reached a vote. Senator Marsha Blackburn, a Republican who had cosponsored an earlier compromise version, ultimately abandoned that compromise and co-led the amendment striking the moratorium outright, warning on the record that the provision could allow big technology companies to continue to exploit kids, creators, and conservatives. Senator Ted Cruz, who had championed the moratorium alongside a companion proposal called the SANDBOX Act that would let AI companies petition to suspend federal rules they considered burdensome, vowed to keep bringing some version of preemption back. Florida's governor, no reliable ally of the Democratic states most active on AI regulation, called the moratorium a subsidy to Big Tech that would let the technology companies run wild, a rare moment of the same criticism arriving from opposite ends of the political spectrum at once.

Congress saying no twice didn't end the effort. It moved the fight from the legislature to the executive branch, the same pattern Chapter 17 found the administration reaching for once a preferred outcome couldn't clear Congress on its own. An executive order established an AI Litigation Task Force inside the Justice Department, tasked with suing states directly over their AI laws, and conditioned states' access to broadband infrastructure funds, the same $42 billion BEAD program authorized under the 2021 Infrastructure Law, on states agreeing to walk back regulations the administration considered too burdensome. The Federal Trade Commission was separately directed to determine whether state laws requiring altered AI outputs to prevent algorithmic discrimination, with Colorado's law cited by name, should be treated as compelling a deceptive trade practice under federal law, a maneuver that would let the federal government use consumer protection law to disable state anti-discrimination rules it couldn't preempt through ordinary means.

What the order chose not to fight is at least as revealing as what it targeted. The executive order explicitly carved out state laws governing child safety and, notably, state authority over AI compute and data center infrastructure, the exact category of regulation that towns and states are beginning to exercise in response to the Broadview, Stratos, and Spartanburg pattern. Those two categories happen to be the ones with the least organized industry opposition and the most bipartisan political support behind them. What the administration chose to spend its political capital fighting was frontier model safety disclosure and algorithmic discrimination liability, the categories of regulation most likely to constrain the labs whose success the government now has a growing financial stake in.

The result, six months in, is neither the sweeping federal preemption the administration wanted nor an unregulated status quo. States enacted 109 AI laws and 28 separate data center statutes in the first half of 2026 alone, according to New York University's tracking, despite the litigation task force and the funding threats. Fewer states are legislating overall than at the same point in 2025, but the ones still moving have converged precisely on the categories the executive order chose not to contest, child safety, data centers, ordinary consumer protection, while the more ambitious algorithmic accountability regimes that states like Colorado pioneered are being quietly narrowed. Illinois passed a law requiring independent third-party audits of frontier developers' internal safety frameworks, building on disclosure requirements California and New York had already put in place.

A handful of states, both Republican- and Democratic-led, from Alabama to Washington, passed new restrictions on how health insurers can use AI in coverage decisions, with Connecticut enacting a ban. Nebraska, which introduced its own version, moved to limit AI-driven dynamic pricing, evidence that at least some of this legislative activity is settling into genuinely cross-partisan consumer protection territory rather than a partisan fight. Colorado's comprehensive AI Act, once the most far-reaching state law of its kind, was rewritten in May after a year of industry pressure and repeated delays to its own effective date into a considerably narrower measure focused only on automated decision-making, stripped of the risk-management and impact-assessment requirements the original law had required.

A bipartisan federal alternative, the Great American AI Act, circulates as a discussion draft introduced by Representatives Jay Obernolte and Lori Trahan, offering transparency reporting, safety-incident disclosure, and whistleblower protections roughly comparable to what California and New York already require of frontier developers but with no guarantee of passage and no fixed timeline. The pattern here is not captured in a straightforward conflict of financial interest. It's closer to a patchwork holding on by inertia, surviving less because the institutions built to produce it are functioning well than because the administration's preferred alternative kept losing votes it needed to win outright.

What Governance Actually Requires

Put the surveillance apparatus and the regulatory fight next to each other and the shape of the problem comes into focus. The same government building the legal infrastructure to preempt state safety rules is, per the *Wired* reporting, simultaneously building the technical infrastructure to monitor its own citizens' opinions about the industry those rules would apply to. Neither fact alone would be alarming. Together, they describe a state whose capacity to watch its own population has expanded faster than its willingness to be watched in return, the exact asymmetry Chapter 1 predicted for any resource that becomes essential enough to attract this much simultaneous financial and political interest.

None of this means the state-level patchwork surviving into 2026 is adequate, and this book shouldn't pretend otherwise for the sake of a tidy ending. A response built out of a 109 separate state laws, an unresolved federal bill, an executive order tied up in litigation, and a surveillance apparatus operating well ahead of any of them is not a governance framework. It's the accumulated residue of several different fights nobody has finished having yet.

The European Union's AI Act, held up elsewhere as the more comprehensive alternative, delayed its own high-risk provisions from 2026 into 2027 after the European Commission itself proposed the postponement, a reminder that even the jurisdiction most committed on paper to getting ahead of this problem is, in practice, still behind it. If the world's most deliberate attempt at comprehensive AI regulation needed an extra year before its hardest provisions took effect, the fifty-state, three-branch improvisation playing out in the US was never going to move faster.

This book's four engines all converge on the same reading of what's happened here, each from its own angle. The Indispensability Trap predicts that something essential enough eventually gets its returns capped, whether through rate regulation, antitrust, or state control, but this chapter has found the trap's access-guarantee mechanism arriving in an unfamiliar shape. Not the state capping the industry, but the state and the industry jointly resisting a cap the states were trying to impose on both of their behalf. Jevons paradox explains why the surveillance capability keeps expanding rather than plateauing once built. Cheaper observation doesn't produce less observation; it produces more of it, applied to more categories of person and belief than the original justification ever contemplated. The Perez framework's installation frenzy is visible in the sheer number of state laws, 109 in six months, racing to get ahead of a technology whose capability keeps moving faster than any legislature's drafting process. The distinction between financial and production capital resolves, in this chapter, into a distinction between institutions actually built to govern and institutions merely built to look like governance until the money sorts itself out. The states did the former. Washington, for most of the first half of 2026, did the latter.

What was left open, whether the institutions entangled in AI's financial success are equipped to police its risks, has a partial answer now, and it isn't reassuring. The institutions most engaged in policing those risks, so far, are the fifty state legislatures the federal government spent 2026 trying to preempt, not the federal apparatus that holds the equity stakes, negotiates the voluntary safety standards, and now runs the surveillance program. That the states did most of the actual governing against active federal resistance, while the entity closest to the industry's balance sheet mostly built tools to watch its critics, is the finding we are left with heading into the final pages. Whether that finding gets corrected in time or simply becomes one more permanent feature of how this build-out ends up governed, is not a question I can answer from inside 2026. It is the question the conclusion has to sit with directly, alongside every other open bet on AI that has yet to find a floor.

Conclusion

The Grid Beneath the Cathedral

> Civilization advances by extending the number of important operations which we can perform without thinking of them.
>
> —Alfred North Whitehead, *An Introduction to Mathematics*, 1911

This book opened in Richland Parish, Louisiana, where Meta's Hyperion campus was rising out of farmland that once grew cotton, and it asked a question that eighteen chapters have now circled from every direction available. When this much capital pours into a single build-out, who ends up holding the value it creates? The historical answer, tested against railroads, telephones, electrification, and the internet, has held with uncomfortable consistency. The builders of essential infrastructure get their returns capped, by regulators, by antitrust lawyers, by strategic capture, or by the brute economics of a commodity. Durable fortunes accrue to whoever treats the finished grid as a free input and builds on top of it. Four times the pattern ran to completion. The fifth time is running now at a scale that makes the first four look like rehearsals.

I want to close by being honest about two things at once, because my entire argument depends on holding them together rather than choosing between them. The first is that, as I said at the outset of this book, I am extremely bullish on AI, full stop. Not cautiously, not with hedges. I believe this technology will prove to be one of the greatest achievements in the history of our species, and I believe the wealth it generates will be measured in numbers we do not yet have good intuitions for. SpaceX's recent public offering contained an addressable market of $28 Trillion dollars. Not to be outdone, Anthropic's filing for their IPO claim an addressable market of $30 Trillion dollars.

The second is that I believe the Indispensability Trap is ironclad, as ironclad as Jevons paradox itself, and that most of that wealth will not land where most of the capital is currently being spent. These are not contradictory positions. They are the same position, examined from two altitudes. The technology succeeds beyond the optimists' dreams, and precisely because it succeeds, intelligence becomes cheap, and cheap is what commodities are, and commodity economics do to the grid what they have done to every grid before it.

The Bull Case, Restated Honestly

In June I published a piece arguing that the AI bull case quietly assumes intelligence gets cheap, and I want to restate that argument here because it is the whole book in miniature. Every trillion-dollar projection for AI's economic impact, every McKinsey curve and sovereign fund thesis and hyperscaler earnings call, rests on the assumption that intelligence will be applied to everything. Every workflow, every document, every supply chain, every customer interaction, every line of code. That assumption is almost certainly correct. But intelligence applied to everything is intelligence that costs almost nothing per application. You cannot have it both ways. Either AI stays expensive and scarce, in which case the trillion-dollar impact projections collapse, or AI becomes ubiquitous and cheap, in which case the impact projections hold and the margins of whoever sells raw intelligence collapse instead. The bull case and the commoditization case are the same case. The optimists have simply not followed their own arithmetic to its destination.

The numbers involved have only grown since the first chapter set the stakes. Combined hyperscaler capital expenditure reached roughly $725 billion in 2026, up 77 percent year over year. McKinsey projects $6.7 trillion in cumulative data center investment by 2030. The International Energy Agency expects data centers to consume 945 terawatt hours annually by the same year. Roughly the electricity consumption of Japan. The fiber build-out of the late 1990s measured at perhaps $90 billion for the core long-haul networks, part of a broader half-trillion-dollar telecom capital wave that vaporized more than a trillion in market value when bandwidth became too cheap to meter. The current build-out spends, in a single year, roughly eight times the entire multiyear fiber investment of the late 1990s. On an annualized basis, more than thirty times what the industry spent per year at the height of the fiber boom. If the pattern holds, and eighteen chapters have found no structural reason it should not, the coming repricing will be proportionate.

A commodity, though, is not the same thing as a worthless product, and this is where the sharpest version of the thesis needs a nuance the earlier chapters only implied. Chapter 10 made the point that intelligence is becoming a differentiated commodity rather than a pure one, gasoline rather than crude, specialty coffee rather than beans, with real but thin premiums available for reliability, safety, and brand. The conclusion has to make the companion point about the sellers.

Nobody Is Taking Up a Collection

Let me say plainly what the thesis does not claim, because the trap is easy to overread. The Indispensability Trap caps the fortune relative to the value created. It does not zero it out, and at the absolute scale of this build-out, the difference matters enormously. When the total addressable spending runs to trillions of dollars a year, even a commoditized, margin-compressed, regulated grid layer generates fortunes that would have made Vanderbilt blink. The railroads made real money for decades even as regulation closed in. AT&T, under the 1913 Kingsbury Commitment and the consent decrees that followed, remained one of the most profitable enterprises on earth for half a century. It simply captured a fraction of the value the telephone network created for everyone who built businesses on top of it. Nobody needs to take up a collection for the hyperscalers. They will be enormous, profitable, essential enterprises long after the margin has migrated up the stack, and anyone reading this book as a prediction of their ruin has read it wrong.

The exception, and it is the exception that has destroyed infrastructure builders in every single wave covered, is financial engineering. Samuel Insull did not lose his empire because electricity commoditized. He lost it because he stacked holding companies on holding companies, pyramided debt against assets whose returns were already being squeezed and left himself no cushion when the financing environment turned. The fiber operators did not die of cheap bandwidth alone. They died of cheap bandwidth plus balance sheets built for expensive bandwidth. The trap compresses margins. Leverage converts compressed margins into insolvency.

That is the mechanism, and it is why the single most important number in Chapter 15's company-by-company map is not any firm's model quality or market share but Oracle's capex-to-revenue ratio of 83 percent. That ratio drove two of the three major rating agencies to a negative outlook. A roughly fivefold widening in its cost of default insurance, from about forty basis points in early 2025 to nearly two hundred by March of this year. Microsoft continues to fund its build-out largely from operating cash flow, but Alphabet, Meta, and above all Amazon have moved decisively toward debt. By 2026 the group is spending roughly ninety cents on capex for every dollar of operating cash flow, up from about sixty-five cents a year earlier, and Amazon's quarterly capex now exceeds its operating cash flow outright. Oracle is borrowing against a future that must arrive on schedule. The same build-out, the same commoditization pressure, and two entirely different exposures to it. Whether Oracle becomes this wave's Insull is one of the cleanest verifiable bets left open, and readers checking back in three years will not need my help scoring it.

The Two Best Arguments Against This Book

An argument is only as strong as the best case against it, so I want to close the analytical half of this conclusion by engaging on the thoughts of two serious thinkers who are extremely bullish on AI infrastructure, not the weakest versions of their positions but the strongest.

Daniel Newman is my friend and partner, and he is the founder and CEO of Futurum Group. Newman's case has been made across LinkedIn, CNBC, and most of the financial news channels. His case is built on the premise that the demand curve for intelligence is so steep and so durable that the build-out is rational at almost any scale, that enterprises will pay sustained premiums for frontier capability. That the capex will be vindicated by workloads we have barely begun to imagine. On the demand side, I think Newman is simply right, and I have never argued otherwise.

Where the argument breaks for me is the assumption that demand for a product protects the margins of its producers. Demand for bandwidth in 2003 was growing at rates that made every fiber prospectus look conservative, and bandwidth prices fell 90 percent anyway. Because in a commodity market, demand growth and price collapse are not opposites, they are the same phenomenon. Jevons wrote it down in 1865: Efficiency gains do not reduce consumption of a resource, they explode it, and the explosion comes precisely through falling prices. Newman's demand curve is real. It is also the mechanism of the margin compression he believes the demand curve will prevent. His bull case confirms the thesis; it does not refute it.

Gennaro Cuofano who writes extensively on his Substack, *The Business Engineer*, makes his case in a subtler fashion, and it deserves a full treatment. In a June 2026 essay called "AI & The Dynamo Doctrine," Cuofano lays out a substrate lineage that is immediately recognizable. Electrons into bits, bits into tokens, each industrial age defined by a machine that consumes one substrate and emits another, with the emitted substrate, not the machine, reshaping the economy. He argues AI is not like electricity but on the same curve as electricity, with the same generational capital requirements, the same regulatory eventualities, and, in his own words, the same commodity endpoint. On all of this we agree completely, and I want to acknowledge the convergence plainly rather than pretend I arrived at the dynamo framing alone. Where we diverge is the conclusion he draws. Cuofano is bullish on the factories themselves. He argues the build-out is underbuilt rather than overbuilt, that the shift from a billion humans consuming intelligence through chat windows to a hundred billion agents consuming it continuously will vindicate every gigawatt. He cites asset-level economics, a gigawatt of AI factory capacity costing roughly $50 billion and emitting $300 to $400 billion in annual token output, that make the landlord position look like the opportunity of the century.

But read Cuofano's stack analysis and a different picture emerges from inside his argument. He breaks the AI economy into nine layers rather than the standard five, and his own accounting of where the money settles is explicit. Capital concentrates in energy, foundries, and compute capacity, while margin concentrates in silicon design, networking, foundation models, agentic orchestration, and distribution. Capital on the ground floor; margin in the upper floors. That is not a rebuttal of the Indispensability Trap. That is the Indispensability Trap stated by an infrastructure bull doing honest arithmetic. His asset-level token economics describe what a gigawatt can emit at today's prices, and today's prices are exactly what Jevons and Cuofano's commodity endpoint promise will not survive the demand explosion he forecasts. Like Newman, his bullishness is about aggregate volume, and about volume he is very likely right. This book's thesis was never about volume. It is about where, within a volume that enormous, the durable margin comes to rest, and on that question Cuofano's nine layers and this book's eighteen chapters give the same answer.

One more detail from his essay arrives at Chapter 13's conclusion from a completely independent direction. Cuofano treats governance not as a layer of the stack but as a control plane cutting perpendicularly through every layer. He observes that a single jurisdictional directive can now switch off a frontier model for every customer on the planet in seventy-two hours. He is describing, in the abstract, exactly what the Fable and Mythos export controls demonstrated in the concrete. That the most advanced products of this build-out are subject to a kill switch held by the state, which is strategic capture in its purest form arriving right on the schedule the trap predicts. When the bulls and the skeptics start describing the same mechanism in the same language, the mechanism is probably real.

The Four Engines, One Last Time

Every chapter has run on the same four analytical engines, and they can be stated together now.

The first is the Indispensability Trap itself. Whatever becomes essential gets its returns capped, through rate regulation, antitrust, strategic capture, or state takeover. The AI build-out has already met the trap in three of its four forms, with export controls, equity stakes, and litigation task forces arriving before the technology is a decade old.

The second is Jevons paradox. Cheaper intelligence will not mean less spending on intelligence but vastly more, which is why the build-out is rational in aggregate even as it ruins particular builders.

The third is the Perez cycle. We are deep in the installation phase, financial capital in command, and the amended version proposed in Chapter 6 still holds. Whether installation ends in a crash or a grind depends on financing structure. That is why the migration of hyperscaler funding from cash flow to debt, $108 billion and more in a single year, is the single most important trend to watch between this book's publication and its inevitable revision.

The fourth is the distinction between financial and production capital. The fortunes that last will belong, as they always have, to whoever uses the cheap output of the grid to reorganize actual work: Paul David's factories rebuilt around the electric motor, the SaaS economy built on commodity cloud, and now whatever gets built next on commodity intelligence, most of which does not yet exist and will not be built by the companies spending the $725 billion.

That is the cold-eyed half of this conclusion, and I will stand behind every sentence of it. But if I ended the book there, I would be committing the same error I have accused the bulls of committing, following the arithmetic carefully while ignoring what the arithmetic is for. So let me pull back to another altitude.

The Cathedral and the Mirror

Last summer I wrote a piece about what is actually driving the people at the top of this build-out, because I had become convinced the standard explanations—market share, shareholder value, competitive positioning—could not account for the behavior I was watching. Men who have already won every game capitalism offers do not mortgage their companies' balance sheets for a technology whose economics they cannot fully model because they want a slightly larger market cap. Something older is at work.

The medieval cathedral builders spent sums their societies could barely afford on structures whose completion they would not live to see because they were building for something beyond the ledger. I wrote then that the AI labs are building cathedrals, for the machines, for themselves, and maybe for their souls, and a year of reporting since has only hardened the view. The pursuit of superintelligence is shot through with ego and with something that looks very much like a pursuit of immortality, the founder's name attached to the thing that outlasts every founder.

There is a still older frame that fits, one I have circled in several pieces this year without quite saying it as directly as I will say it here. The quest for superintelligence is a god-like ambition in the most literal sense available to a secular industry. We are attempting to create consciousness, if not in our own image, then certainly with our own hands and our own minds. Whatever one believes about whether these systems will ever be conscious, and I claim no ability to settle that question, the ambition itself is the ambition of Genesis run in reverse. The created reaching for the power of the creator. Humanity has told itself cautionary stories about exactly this reach for as long as it has told stories at all, from Prometheus to the golem of Prague to Frankenstein. It says something that the people building these systems increasingly reach for those stories themselves rather than for the language of quarterly guidance.

It says even more that the world's oldest institutions have noticed. This spring I wrote about a convergence that would have sounded like satire five years ago. The Vatican's release of Pope Leo XIV's first encyclical, *Magnifica Humanitas*, personally presented by the pope with Anthropic cofounder Chris Olah in the room, even as rabbis, Buddhist monks, and AI lab researchers were independently arriving at strikingly similar conclusions in the same season, asking the same questions about what machine efficiency does to human flourishing. A few weeks earlier I had written about the labs themselves turning to clergy, academics, and ethicists for help with questions their own disciplines could not frame. I argued that the most important question in those rooms was never whether AI has a soul. It is whether the people building it remember that they do. When the Pope, the rabbis, the monks, and the AI labs are all having the same conversation, that is not a curiosity for the religion page. That is civilization noticing that the stakes of this build-out run past the balance sheet, past the grid, past every question this book has spent eighteen chapters trying to answer.

I hold both of these truths without strain, and I think the reader should too. This may genuinely be one of the greatest things humanity has ever attempted. The wealth it creates will be real and vast, the science it accelerates will be real, the diseases it helps cure and the knowledge it unlocks will be real. None of that is diminished by the equally real fact that the money will pool somewhere other than where the concrete is being poured. Awe at the achievement and clear eyes about the economics are not competing loyalties. They are what honesty looks like when the subject is this large.

An Unfinished Story

This book goes to press in the middle of its own subject. The four historical waves it rests on took decades to run to completion. The fifth is perhaps four years into a cycle whose installation phase alone will outlast this edition. Chapter 15 left clear markers against every major position on the board. Chapter 13's volunteered-capture episode was still developing as the ink dried. Oracle's balance sheet, the preemption fight, the Cuofano demand curve, the open-weights price floor, every one of these is a live experiment whose results will come in on their own schedule, indifferent to publishing deadlines. I intend to keep updating this story as it unfolds, in my columns and, if the material demands it, in another book, because the honest way to end an argument about a pattern still in motion is to specify what would prove it wrong and then keep watching.

But the pattern itself, I am confident, will hold, because it has held four times across two centuries for reasons that have nothing to do with any particular technology and everything to do with what societies eventually do to anything they cannot live without. The grid gets built. The grid becomes essential. The essential gets captured, capped, and commoditized. The value flows up the stack, to the builders of appliances, to the reorganizers of work, to whoever figures out what cheap intelligence makes possible that expensive intelligence never could. It will take place somewhere in Richland Parish, and in Broadview, and in towns like Spartanburg that have already said no. In a thousand others that never asked to host the fifth great build-out, that future is being poured into concrete right now, financed by a handful of companies and a handful of men making the largest infrastructure bet in the history of money.

They are reaching for something extraordinary, and some of what they build will belong to all of us, the way the railroads and the wires and the fiber eventually did, long after the fortunes made and lost in building them were forgotten. Whether they intend it or not, that is how every one of these stories has ended. One small step for men, perhaps even for a few men. But a giant leap for mankind.

Sources and Further Reading

The complete Sources and Bibliography for this book is available online and searchable at:

https://indispensabilitytrap.com/sources-bibliography.html

What follows is not an academic apparatus. It is a chapter-by-chapter account of where the numbers, quotes, and stories came from, and where a reader who wants to go deeper should go next. Figures for the AI build-out move fast; each is cited to its source as of mid-2026.

Chapter 1: The Trillion-Dollar Bet

The Hyperion figures come from Riley Griffin's reporting in *Bloomberg Businessweek* (May 18, 2026): Meta's planned spend of well over $200 billion, the roughly five-gigawatt draw, the ten gas-fired plants under the expanded March 2026 Entergy agreement, the property tax reductions approaching 80 percent, and the nondisclosure agreements binding state officials. The comparison of five gigawatts to New York City's average electricity draw follows the same convention *Bloomberg* uses.

The Stratos figures for Box Elder County, Utah, the forty thousand acres, up to nine gigawatts at full build-out, and $35 to $50 billion of initial investment, come from the project's public materials at thestratosproject.com. The projection of one to two thousand permanent jobs comes from the Utah Governor's Office and Box Elder County fact sheets. Readers should weigh that those are proponents' numbers.

The Broadview, Montana, campus is the Big Sky Digital Infrastructure Campus. Its permanent jobs figure has not been publicly disclosed; the image of a workforce seated in the town's only restaurant is mine.

The cumulative build-out estimate of roughly $7.6 trillion through 2031 comes from the Goldman Sachs Global Institute report "Tracking Trillions: The Assumptions Shaping the Scale of the AI Build-Out" (Lee and Greenwillbaum, May 1, 2026).

The hyperscaler capital expenditure figures, roughly $725 billion in 2026, up more than 70 percent from a 2025 base of about $410 billion, aggregate the four companies' own guidance at the top of their announced ranges.

The electricity projection, data center consumption roughly doubling from 415 to 945 terawatt hours by 2030, with the comparison to Japan's annual usage, comes from the International Energy Agency's "Energy and AI" report (April 10, 2025).

The epigraph, and the installation-and-deployment vocabulary used throughout this book, come from Carlota Perez, *Technological Revolutions and Financial Capital: The Dynamics of Bubbles and Golden Ages* (2002).

Chapter 2: Grid and Appliance

The epigraph comes from *The Gilded Age* (1873), the novel Mark Twain cowrote with Charles Dudley Warner; the line is conventionally attributed to Twain.

This chapter builds its model from the cases documented elsewhere in the book, and its claims are sourced where those cases are told in full: the railroad material under Chapter 3, the General Electric story under Chapter 5, and the three recurring forces, the Indispensability Trap, Jevons paradox, and the Perez installation-and-deployment cycle, under Chapter 1. Vanderbilt's fortune, the model's hardest test case, is sourced under Chapter 3.

Chapter 3: In the Railroad Business, Not the Transportation Business

The chapter's frame and its title come from Theodore Levitt's "Marketing Myopia," *Harvard Business Review* (1960).

The rail mileage figures, roughly 9,000 miles in 1850, fifty thousand by 1870, 160,000 by 1890, and past 200,000 by the turn of the century, follow the standard series used by the Association of American Railroads and reproduced in most economic history surveys.

The land-grant accounting, roughly 175 million acres authorized against about 130 million actually patented, comes from Paul Wallace Gates, *History of Public Land Law Development* (1968).

The Crédit Mobilier story, Thomas Durant, the Ames brothers, and the roughly $94 million charged for $50 million of work, follows Maury Klein's *Union Pacific*, volume one.

The Day of Two Noons and the *Indianapolis Daily Sentinel*'s complaint that "the sun is no longer to boss the job" (November 21, 1883) come from Ian Bartky, *Selling the True Time* (2000).

The railroads' invention of the managerial corporation is Alfred Chandler's argument, made across *The Visible Hand* and his earlier work.

The 1893 receivership figures, roughly a quarter of national mileage, follow Richard White, *Railroaded* (2011).

The regulatory arc runs through *Munn v. Illinois* (1877), the Wabash decision (1886), and the Interstate Commerce Act, signed February 4, 1887.

Vanderbilt's fortune, about $105 million at his death in 1877, one eighty-seventh of the national economy, or roughly one of every $20 then in circulation, follows T.J. Stiles, *The First Tycoon* (2009).

The Carnegie material, the $480 million price settled on a slip of paper, Carnegie's share of better than $225 million in fifty-year gold bonds, US Steel's $1.4 billion capitalization, and Morgan's "richest man in the world" remark, follows David Nasaw, *Andrew Carnegie* (2006).

The rebate and drawback mechanics, the South Improvement Company scheme, Rockefeller's roughly $900 million stake in 1913, and the September 1916 billionaire milestone follow Ron Chernow's *Titan* (1998).

The Sears and Rosenwald material, including the 4,977 schools built in partnership with Booker T. Washington between 1912 and 1932, follows Peter Ascoli's *Julius Rosenwald* (2006).

The junction-city argument is William Cronon's, from *Nature's Metropolis* (1991).

The octopus is Frank Norris's, from his 1901 novel of the same name.

Chapter 4: The Monopoly That Asked to Be Regulated

Bell's patent, No. 174,465, was granted March 7, 1876, on an application filed February 14, the same day Elisha Gray's caveat arrived at the Patent Office; the Library of Congress and the Franklin Institute case files carry the details, and the exact interval between the two filings has been argued about for a century and a half. The first commercial exchange opened in New Haven on January 28, 1878, with twenty-one subscribers.

The Vail biography, the Railway Mail years, the Fast Mail, the 1878 hiring by Hubbard, the 1887 resignation, the Argentina interlude, and the 1907 return after the failed Morgan bond sale, follows the standard accounts, principally American Heritage's "Breaking the Connection" and the utility-history literature. The roughly six thousand independent telephone companies that rose after the patents lapsed in 1893 and 1894, and Bell's competitive tactics against them, follow the same literature.

The Kingsbury Commitment is the letter of December 19, 1913, from AT&T vice president Nathan Kingsbury to Attorney General McReynolds; its three terms are quoted from the letter itself. The Willis Graham Act (1921) and its consequence, the Interstate Commerce Commission approving 271 of AT&T's 274 subsequent acquisitions, follow the legislative record. The Communications Act of 1934's universal service language paraphrases 47 U.S.C. §151, and the Supreme Court's characterization of implicit subsidies is paraphrased from *FCC v. Consumers' Research* (2025).

The transcontinental line opened January 25, 1915, Bell in New York and Watson in San Francisco, the last pole at Wendover on the Utah-Nevada line. Bell Labs was founded January 1, 1925; its inventions and honors are documented in the standard histories. The 1956 consent decree signed January 24, 1956, and its royalty-free licensing of the patents AT&T then held, including the transistor, follows the Federal Judicial Center's account and the economics literature on the decree, particularly Watzinger and Schnitzer's work.

The breakup runs from the 1974 Justice Department suit through the January 8, 1982, settlement before Judge Harold Greene to the January 1, 1984, divestiture of the seven regional operating companies, per the Federal Judicial Center. The reassembly: SBC acquired AT&T Corporation in 2005 for roughly $16 billion and took the name; Bell Atlantic became Verizon by way of NYNEX and GTE; and AT&T's purchase of Lumen's consumer fiber business, assets tracing back through Qwest to US West, closed in early 2026, per the companies' own announcements. British Telecom's November 1984 privatization, roughly £3.6 billion and the largest stock flotation to that date, follows the Institute for Government's study.

Chapter 5: The Man Who Asked to Be Regulated

The Kemmler execution took place August 6, 1890, at Auburn Prison; Westinghouse's "they could have done better with an axe" is his historically attested reaction. The War of the Currents material, Edison's secret funding of Harold Brown's demonstration electrocutions, and the campaign to have the chair run on alternating current, follows the Rutgers Edison Papers Project. The same archive, along with *Smithsonian Magazine*, supports the chapter's correction of the Topsy myth: the elephant was killed at Luna Park in January 1903, years after the currents war ended, and not on Edison's order.

Westinghouse lit the World's Columbian Exposition in 1893, roughly 27 million visitors, and Niagara power reached Buffalo on November 16, 1896.

The Insull biography follows the Loyola University archives, the American Business History Center's account, and the utility-history literature. "Massing production" is Insull's term, preserved in his published speeches (*Central-Station Electric Service*, 1914). The regulated monopoly philosophy dates to his June 1898 address to the National Electric Light Association; the phrase that titles the chapter's epigraph is from his 1910 address of that name, the text of which survives.

The fall: Middle West Utilities entered receivership in April 1932, the $10 million loan against a $3 billion empire follows the contemporaneous valuations, and the three trials of 1934 and 1935 ended in acquittal, dismissal, and directed verdict, per *TIME*'s and the *New York Times*'s contemporaneous coverage. Insull died in a Paris Métro station on July 16, 1938. The legislative aftermath is a matter of statute: the Public Utility Holding Company Act (August 26, 1935), the Rural Electrification Administration (May 1935), and the Rural Electrification Act (May 1936).

Ford's moving assembly line began at Highland Park on October 7, 1913, per Ford's corporate history; the assembly-time and price figures are the company's standard numbers. The Wisconsin Public Utilities Law of 1907 is the model regulatory compact the chapter describes.

Chapter 6: The Roads Were Already Built

The Interliant story, the Houston meeting, and the Qwest negotiations are firsthand; I was in the rooms. The reported history around them locks as follows.

The Telecommunications Act was signed February 8, 1996. Global Crossing filed for bankruptcy on January 28, 2002, with roughly $12.4 billion in debt and a hundred thousand miles of cable; the *Los Angeles Times* coverage of January 29, 2002, remains the cleanest contemporaneous account. WorldCom followed on July 21, 2002, with $41 billion in debt and an accounting fraud initially measured at $3.8 billion; Bernie Ebbers went to prison for it.

Qwest's descent from the Southern Pacific Railroad, with more than two-thirds of its fiber running along old railroad rights-of-way, is documented in *Fortune* (July 5, 1999) and the *New York Times* (May 6, 2000). Sprint's name really was an acronym: Southern Pacific Railroad Internal Networking Telecommunications. Corning's fall from over $100 a share to about $2 matches the recorded band (above $113 in September 2000, near $1.10 at the October 2002 bottom).

The cloud and container timeline runs on the industry's records: VMware Workstation (May 15, 1999), ESX Server (March 2001), Intel VT-x (November 2005) and AMD-V (May 2006), Amazon S3 (March 14, 2006) and EC2 (August 24, 2006, per the AWS launch releases), Docker's debut at PyCon (March 20, 2013, Solomon Hykes), and Kubernetes (first commit June 6, 2014; version 1.0 on July 21, 2015).

The Google search-monopoly ruling came down August 5, 2024; the observation that the Justice Department used the Microsoft case as its roadmap follows the contemporaneous legal analysis, including CNBC's (August 7, 2024).

Chapter 7: The Fork in the Road

The epigraph is Andrew Grove's from *Only the Paranoid Survive* (1996). The Goldman Sachs $7.6 trillion figure is sourced under Chapter 1.

The SpaceX absorption of xAI closed as an all-stock merger in February 2026, per Bloomberg and CNN. The Google-SpaceX arrangement, $920 million a month for roughly 110,000 GPUs, October 2026 through June 2029, north of $30 billion total, cancelable on ninety days' notice, was disclosed in SpaceX's pre-IPO filings and reported by CNBC in June 2026. The Anthropic lease of Colossus 1 at $1.25 billion a month appears in the same prospectus; Musk's subsequent public description of it as a six-month lease is noted in the text because the two framings genuinely conflict. The Google-Anthropic TPU arrangement, $40 billion for up to a million Ironwood chips, follows Anthropic's partnership announcement and the reporting around it.

The 70 percent compute concentration statement was made by U.K. Technology Secretary Liz Kendall in April 2026 and repeated at London Tech Week that June, per the government's release and *Guardian* coverage. The EU's AI Gigafactories carry a €20 billion InvestAI commitment for up to five facilities, per the European Commission. Canada's sovereign compute initiative launched in April 2026, with the broader national strategy following in June. Roughly ninety countries and organizations signed the New Delhi Declaration on AI sovereignty.

The Fable and Mythos timeline, released in June 2026, suspended within days under Commerce Department controls, restored July 1, is sourced fully under Chapter 13; CSIS's analysis noted it was the first use of the Export Control Reform Act's worldwide license requirement mechanism against an emerging technology company. Mistral's licensing history is sourced under Chapter 10. Brad Feld's rule of three comes from my long acquaintance with him.

Chapter 8: Landlord, Tenant, Rival

The epigraph is John Kenneth Galbraith's from *A Short History of Financial Euphoria* (1990).

Microsoft's quarterly figures, $37.5 billion in capital expenditure, up 66 percent, come from the company's fiscal 2026 second-quarter earnings. The restructured OpenAI partnership, a nonexclusive license running through 2032, was announced by Microsoft itself on April 27, 2026, and reported by Reuters the same day. Copilot Cowork launched March 9, 2026, per Reuters; the MAI family of in-house models (MAI-Voice-1, MAI-Transcribe-1, MAI-Image-2) shipped April 2, 2026, weeks before the partnership rewrite.

Amazon's $50 billion investment in OpenAI, with OpenAI's two-gigawatt Trainium commitment, was announced February 27, 2026, per Reuters, CNBC, and OpenAI's statement. The mirror image Anthropic arrangement of April 20, 2026, up to $25 billion in new investment atop the $8 billion since 2023, alongside Anthropic's commitment of more than $100 billion in AWS spending and up to five gigawatts of Trainium, follows the companies' announcements and GeekWire's filings analysis.

Anthropic's revenue trajectory, roughly $1 billion at the start of 2025 to $5 billion by August, $9 billion by year-end, past $30 billion by April 2026, follows the company's disclosures. Google's token processing growth, from 480 trillion monthly tokens in May 2025 to 3.2 quadrillion by May 2026, comes from Sundar Pichai's successive I/O keynotes and Alphabet's earnings disclosures. OpenAI's projected $14 billion loss for 2026 against $100 billion in projected 2029 revenue follows the widely reported internal projections.

Chapter 9: The Substrate

The epigraph is Morris Chang's widely reported remark at TSMC's Arizona tool-in ceremony in December 2022.

TSMC's 2025 results, roughly NT$3.6 trillion in revenue, up 35.9 percent, with margins above 45 percent, come from the company's January 2026 earnings disclosures. Its share of leading-edge fabrication (above 90 percent at the most advanced nodes) and of pure-play foundry revenue overall (roughly 70 percent) follow TrendForce and Counterpoint tracking, as does Samsung's roughly 7 percent share against a 2021 peak in the high teens.

The $165 billion US investment commitment is TSMC's, announced March 2025. The Arizona fab timelines come from the company's January 2026 earnings call. The CoWoS advanced packaging capacity trajectory, from 35,000 wafers a month in 2023 toward 120,000 to 130,000 by the end of 2026, follows TrendForce and industry capacity trackers; C.C. Wei's statement that the capacity is sold out through 2027 was made at TSMC's annual shareholder meeting on June 4, 2026.

The Intel material: the US Treasury's 9.9 percent stake for $8.9 billion at $20.47 a share was announced August 22, 2025; Nvidia's $5 billion common-stock purchase, roughly 4 percent of the company, is documented in the FTC-approved deal filings; the 18A yield improvements of roughly 7 percent a month follow TrendForce reporting through late 2025 and into 2026.

The 15 percent remittance on Nvidia and AMD China chip revenue was announced August 10, 2025 (Reuters). Nvidia's China market share falling to zero is Jensen Huang's disclosure, made on the company's May 2026 earnings call. China's target of 80 percent chip self-sufficiency by 2030 follows TrendForce and Nikkei Asia reporting.

The $2.5 trillion estimate for the economic cost of a Taiwan disruption is the figure used across the policy literature and cited in *New York Times* and *Foreign Policy* coverage. The December 2025 exercises around Taiwan were assessed as the largest on record by the Institute for the Study of War and Reuters. The 2027 readiness benchmark is the Pentagon-referenced directive from its annual China Military Power Report. The 22 percent conflict probability is Manifold Markets' early-2026 pricing, attributed as such in the text. The xLight figures, a $150 million CHIPS Act commitment plus a $350 million Series B, follow Manufacturing Dive and industry reporting as of mid-2026.

Chapter 10: The Free Model

The epigraph is Stewart Brand's, from the first Hackers' Conference in 1984.

The gap between the best closed and best open models, 3.3 percentage points as of March 2026, up from half a point in August 2024, and the composition of the top ten leaderboard come from Stanford's 2026 AI Index. OpenAI's gpt-oss-120b shipped in August 2025 under Apache 2.0, with benchmark results from the company's model card.

The training-cost comparison, DeepSeek V3 at roughly 2.8 million GPU hours against Llama 3 405B at roughly 30.8 million, comes from DeepSeek's technical paper and Meta's published documentation. DeepSeek V4-Pro's 80.6 percent on SWE-bench Verified comes from its April 2026 release materials and the benchmark trackers. GLM-5.2's results against GPT-5.5 on long-horizon coding, at roughly one-sixth the cost, follow Zhipu's June 2026 release and the independent evaluations published alongside it.

The July 2026 wave is sourced to the releases themselves and the contemporaneous coverage: Kimi K3 (Moonshot AI, July 16, 2026; 2.8 trillion parameters, million-token context, weights committed for release by month's end), its placement on the Artificial Analysis Intelligence Index above one of the closed frontier flagships, and the Arena blind testing preference for front-end coding, per Artificial Analysis's published rankings and the reporting at Axios, CNBC, and the South China Morning Post; and Qwen 3.8 tMax (Alibaba, July 19, 2026; 2.4 trillion parameters), with Alibaba's capability claim attributed as the company's. The leaderboard-composition observation, close to half of the top dozen models being Chinese and most of those open weight, is hedged in the text precisely because the rankings disagree with one another; it holds across the major composite indexes as of late July. The calls from American AI companies for regulation of Chinese open models follow the industry press of the same week, including HPCwire's coverage. The framing of China's open strategy, and the line about which contest each country is running, draws on my own contemporaneous essay on China's open-weight strategy, written as the July wave broke. Token prices moved throughout; the figures here reflect published provider rates as of late July 2026.

The Mistral material, Large 3 released December 2025 as a 675-billion-parameter mixture-of-experts model trained on three thousand Nvidia H200s under Apache 2.0, the ARR growth from roughly $20 million to roughly $400 million, and the $13.8 billion valuation, follows the company's launch materials, model card, and contemporaneous reporting.

Chinese open-weight providers passing 45 percent of OpenRouter traffic in the second quarter of 2026, and MiMo V2 Pro's roughly 4.8 trillion tokens a week, follow the platform's published usage data and industry analyses. Alibaba's disclosure that the Qwen family passed 700 million cumulative downloads came in January 2026. Cursor's use of Kimi K2.5 and Reflection AI's $8 billion valuation (a $2 billion round in October 2025) are on the record. MiniMax's license shift from MIT to noncommercial is documented in the licensing trackers.

The price collapse figures, declines of 9x to 900x per year depending on capability tier, and GPT-3-level output falling from $60 per million tokens in late

2021 to six cents by early 2026, come from Epoch AI's tracking and Andreessen Horowitz's "LLMflation" analysis, which coined the term. OpenAI's roughly $13 billion in late-2025 run-rate revenue against losses estimated at $8 to $12 billion follows the reporting ahead of its expected public offering, including my own at Techstrong.ai. Alphabet's capital expenditure guidance of $180 to $190 billion for 2026 comes from its first-quarter 2026 earnings call. Martin Casado's 80 percent figure is from his own public statements, corroborated in the US-China Economic and Security Review Commission's March 2026 report.

Chapter 11: China's Two Loops

DeepSeek R1 was released January 20, 2025, and its app topped the US App Store within the week; Nvidia's single-day loss of nearly $600 billion on January 27 was the largest in US market history, per CNBC and NBC. Marc Andreessen's "Sputnik moment" post is dated January 26, 2025. The "diffusion-innovation feedback loop" framing, and the chapter's title concept, come from the US-China Economic and Security Review Commission's March 2026 report *Two Loops: How China's Open AI Strategy Reinforces Its Industrial Dominance*, by Ngor Luong.

Li Qiang's July 2025 WAIC keynote, the thirteen-point Global AI Governance Action Plan, and the proposed World AI Cooperation Organization follow Xinhua's coverage and Carnegie's analysis. The AI+ Initiative was formally approved by the State Council in August 2025. The 15th Five-Year Plan's four named AI frontiers follow the plenum documents and Trivium China's analysis. China's daily token consumption, 140 trillion by March 2026 against roughly 100 billion at the start of 2024, is the National Data Administration's figure.

The diffusion evidence: Chinese models at 61 percent of OpenRouter token consumption in February 2026, China passing the United States in trailing Hugging Face downloads, MiniMax's Hong Kong debut doubling on its first day in January 2026, and Singapore's national SEA-LION program moving to a Qwen-based architecture, all follow the platform data and contemporaneous reporting cited in the text.

Anthropic's February 2026 disclosure that DeepSeek, Moonshot, and MiniMax had used Claude output to train competing models is the company's, covering roughly 24,000 accounts and more than 16 million exchanges. DeepSeek's failed attempt to train R2 on Huawei Ascend hardware follows the *Financial Times*' August 2025 reporting. The government bans (Italy, Taiwan, and the rest), South Korea's finding of data transmission to ByteDance, the Wiz Research exposed database, and the Feroot Security identification of China Mobile-linked code follow the coverage cited in the text, including my own reporting at Techstrong.ai. Article 7 of the 2017 National Intelligence Law is paraphrased from the standard translation. Beijing's block of Meta's acquisition of Manus came April 27, 2026, per Reuters and Bloomberg.

Chapter 12: Sovereign Ground

Erik Voorhees's description of Venice AI's architecture is quoted from his own writing. The Council of the European Union amended the EuroHPC Joint Undertaking's mandate in January 2026. The UK sovereign fund arithmetic, £500 million against OpenAI's $852 billion valuation, roughly 0.08 percent, follows *Computerworld*'s April 2026 analysis; the candid assessment of the ambition is Amine Abidi's, a senior partner in Kearney's digital practice. The Lumen Sovereign backer coalition (BT, Lloyds, NatWest, LSEG, BAE Systems, Babcock, Thales, PwC, the Alan Turing Institute) follows the venture's materials.

The Gulf material: G42's divestment of its Chinese holdings, announced February 2024; HUMAIN as a subsidiary of Saudi Arabia's Public Investment Fund, with Aramco entering as minority partner under an October 2025 term sheet; Tareq Amin's full-stack ambitions, paraphrased from his public statements; and the 18,000-unit GB300 shipment, announced May 2025 and confirmed delivered that December following US export authorization. The Falcon models are the Technology Innovation Institute's. The April 2026 incident at Oracle's Dubai facility follows the contemporaneous reporting.

The sovereignty framework the chapter engages is the Tony Blair Institute for Global Change's January 2026 report *Sovereignty in the Age of AI: Strategic Choices, Structural Dependencies*. The IndiaAI Mission figures, the GPU pool expanding from roughly 38,000 toward 100,000, BharatGen's Param2 (17 billion parameters, 22 official languages), and Sarvam-105B, follow the mission's announcements and summit coverage.

Chapter 13: The Switch

This chapter was source-verified against contemporaneous reporting rather than a retrospective literature, because the events were weeks old as it was written; every date and figure reflects the public record as of mid-2026.

The Fable and Mythos suspension timeline, the Commerce Department letter, the roughly three-week outage, and the July 1 restoration with approved-organization access granted June 26, follows the contemporaneous coverage of the episode and Anthropic's public statements. The named reactions are quoted or characterized from their authors' public commentary: Dean Ball's criticism of the directive's internal contradiction, Gary Marcus's warnings on researcher flight and investor confidence, and the enterprise-consultant framing of capability sovereignty.

The OpenAI equity proposal was reported by the *Financial Times* on July 2, 2026: a 5 percent stake worth roughly $42.6 billion against the $852 billion valuation set in the company's March 2026 funding round, modeled on the Alaska Permanent Fund and pitched by Sam Altman directly to President Trump, Commerce Secretary Lutnick, and Treasury Secretary Bessent. The proposal's antecedents are on the record: OpenAI's April 2026 Public Wealth Fund policy paper, the February 2026 executive order calling for a federal sovereign wealth fund, the Treasury's August 2025 conversion of CHIPS Act grants into a roughly 10 percent Intel stake (sourced fully under Chapter 9), and the 15 percent, later 25 percent on some shipments, remittance on Nvidia and AMD China chip revenue. OpenAI's confidential SEC filing and the SoftBank tranche priced against the same $852 billion figure follow the financial press reporting of the same period.

Senator Sanders's American AI Sovereign Wealth Fund Act, with its one-time 50 percent stock tax on AI companies above $200 million in AI revenue, is characterized from the bill text and Sanders's framing. Steve Bannon's "tip money" dismissal is from his public commentary. Anthropic's universal predistributive capital accounts proposal is from the company's policy paper released the previous month, and its nonparticipation in equity discussions follows the reporting in the days after the *Financial Times* story.

The governance critiques are quoted from their authors: Nat Purser of Public Knowledge on the shareholder-and-regulator conflict, Bruce Schneier's argument for a public AI option, and Tad DeHaven of the Cato Institute on political control through voting shares. The legal analysis of the government's existing equity positions, including the absence of an express authorizing statute, follows the scholarship published after the Intel stake.

The polling: 55 percent of Americans expecting AI to do more harm than good is Quinnipiac; 71 percent saying AI is moving too fast for regulators is YouGov, both from the same season. Governor Newsom's executive order directing California agencies to evaluate resident ownership stakes was signed in May 2026. OpenAI's restricted rollout of its GPT-5.6 models at government request was reported contemporaneously.

Chapter 14: Thirty Days from Going Out of Business

The founding scene in the Denny's booth on Berryessa Road, April 1993, follows Huang's own retelling in his May 2024 Stanford commencement address and CNBC's reporting of the same. The résumés (Huang from LSI Logic; Malachowsky and Priem from Sun Microsystems) and the Sequoia $2 million seed in 1993 followed by the Sequoia and Sierra round in June 1995 follow Nvidia's own corporate timeline pages. The NV1's fatal quadratic-surface design against Microsoft Direct3D's triangles, the RIVA 128's August 1997 launch (a million units in four months, one month of payroll remaining), and the Sega $5 million lifeline from Shoichiro Irimajiri are documented in the IEEE Computer Society's Chasing Pixels series and Tom's Hardware's account.

The IPO is prospectus record: Nasdaq listing January 22, 1999 at $12 per share, roughly $563 million market capitalization. The GeForce 256 (announced August 31, 1999; released October 11) marketed as "the world's first GPU," the November 8, 2006 unveiling of CUDA, the September 30, 2012 AlexNet result (Krizhevsky and Sutskever under Hinton, two GTX 580s), and the August 9, 2016 delivery of the first DGX-1 to OpenAI with Huang's handwritten inscription, are all documented in the IEEE series, contemporaneous trade coverage, Fortune's coverage of the delivery, and Huang's own posts. The Mellanox purchase ($6.9 billion, announced March 11, 2019) follows Nvidia's own press release; the Arm deal's abandonment on February 8, 2022 follows the New York Times and The Guardian.

The market-cap ladder is exchange record: $1 trillion intraday May 30, 2023;

$2 trillion February 23, 2024; $3 trillion June 5, 2024; $4 trillion first close on July 9–10, 2025 (first company ever); $5 trillion October 29, 2025 (first company ever). Reuters and CNBC carried each milestone. The split-adjusted return from the IPO price is high and the chapter's "roughly one million percent" is a round figure; Nasdaq's tracking through late 2024 showed roughly 373,000 percent, doubling into the October 2025 milestone, so the more precise reading is closer to 900,000 percent.

The Q2 FY27 print is Nvidia's own release of August 26, 2026: revenue $96.2 billion (up 106 percent year over year), data-center revenue $89.0 billion, gross margin 75 percent, Q3 guidance in the range described. Kress's guidance on hyperscaler capex (roughly $600 billion in 2025 climbing to $3 to $4 trillion by 2030) tracks CNBC's earnings coverage and Nvidia's GTC 2026 messaging; Huang's $1 trillion Blackwell-plus-Vera-Rubin visibility number follows Yahoo Finance's writeup. The Amazon 2-million-GPU commitment landed the same day and is confirmed in Nvidia's release and Bloomberg's coverage. Market-share estimates (roughly 92 percent in 2023 declining to 80 to 85 percent by 2026) track industry trackers including eeNews Europe and presenc.ai; Google's TPU program is now in its eighth generation (Ironwood, Sunfish, Zebrafish); AMD holds 5 to 7 percent.

The investment book, $42.783 billion in marketable securities, $51.157 billion in non-marketable equity, and roughly $18 billion of additional commitments, comes from Nvidia's 10-Q for the quarter ended July 26, 2026, filed with the SEC on August 26, 2026, and is summarized in Unite.ai's readout. The FY26 $17.5 billion into private funds and infrastructure vehicles and the sixty-seven venture deals in 2025 follow TechCrunch's tracker and CNBC's Q4 FY26 coverage. The specific stakes track cleanly across primary releases: OpenAI's up-to-$100 billion framework for as much as 10 gigawatts of Nvidia systems (Reuters, September 22, 2025; the Wall Street Journal has since reported the framework "on ice," and the chapter's imputed $30 billion current-stake shorthand should be read against that filing gap); Anthropic $10 billion paired with Microsoft's $5 billion; xAI $10 billion rolling into a roughly $21 billion SpaceX position after the all-stock merger; CoreWeave 7 percent plus $6.3 billion of unsold-capacity purchases through 2032; Nebius $2 billion (March 11, 2026); Intel 4 percent for $5 billion (September 18, 2025); Safe Superintelligence $5 billion (July 27, 2026); Naver $1 billion at a 4.5 percent stake; and $2 billion each into Lumentum, Coherent, Marvell, and Synopsys across March 2026. Huang's language about not picking winners and investing in "everybody who's got a good idea" is from his April 15, 2026 conversation with Dwarkesh Patel, reported in CNBC, TheStreet, and Business Insider.

The Groq transaction ($20 billion cash, announced December 24, 2025, structured as a non-exclusive inference-technology licensing agreement with Ross, Madra, and the engineering team joining Nvidia while Groq nominally continues under CFO-turned-CEO Simon Edwards) follows Groq's own blog post, CNBC, the Reuters wire, and CNBC's follow-up reporting on the "fiction of competition" structure. The Warren-Blumenthal letter of March 2026 questioning the deal under the antitrust laws and the FTC's "reverse acquihire" inquiry follow Senator Warren's press release and Bloomberg Law's coverage. The Hugging Face acquisition ($12.93 billion, entered September 2, 2026, disclosed September 3, targeted close in the first half of 2027 pending regulatory clearance) follows TechCrunch, Nvidia's own blog announcement, and the Business Model Analyst readout of Nvidia's 8-K. The 18 million developer, 3 million model, and 1 million application scale figures are Hugging Face's own. Delangue's 2025 refusal of a $500 million Nvidia offer at a $7 billion valuation is documented in Observer's profile and OODA Loop; the security-breach reasoning tied to the July 2026 OpenAI-agent escape into Hugging Face's infrastructure is from TechTimes.

The DSX AI Factories program (launched July 1, 2026) and its credit-support-and-revenue-share model follow Nvidia's own blog announcement, The Futurum Group's analysis, and Moneycontrol. The August 10, 2026 memorandums of understanding with Apollo, BlackRock, Blackstone, Brookfield, Goldman Sachs, and KKR to mobilize more than $500 billion of third-party capital follow Nvidia's investor-relations release, CNBC, and Reuters; Huang's "compute is revenue" framing and David Solomon's "market for credit backed by Nvidia compute" line come out of the same release cycle, with Constellation Research and The Futurum Group carrying the analysis of AI factories as an investable asset class. The $105 billion contingent guarantee tied to OpenAI's 4.25-gigawatt Pike County, Ohio build-out and Nvidia's $1.5 billion equity investment in SB Energy follow CNBC and Reuters, with INDmoney's decomposition the cleanest single accounting of how the $500 billion mobilization target and the $105 billion residual-value ceiling fit together. Bank of America's projection of Nvidia free cash flow exceeding $400 billion across 2026 and 2027 is from BofA's report as summarized on Investing.com. The CoreWeave credit-default-swap widening to roughly 855 basis points, the market-implied five-year default probability of around 50 percent, and the observation that annual interest expense now exceeds 2025 revenue, follow TechTimes's coverage of the CDS move.

The China sequence follows the export-control record. The A100 and H100 license requirement reached Nvidia in late August and early September 2022; the general rule was published October 7, 2022 (effective October 21), and the chapter's "October of 2022" is defensible against the broader-rule date, per the Congressional Research Service report on U.S. Export Controls and China and BIS's own information page. The April 2025 restriction on the China-tuned H20 chip and the disclosed $5.5 billion writedown, the July 15, 2025 reversal after Huang's direct lobbying, the December 8, 2025 authorization of the more capable H200 conditioned on a 25 percent revenue share paid to the U.S. Treasury, and Under Secretary Kessler's July 14, 2026 congressional testimony that H200 shipments to Chinese buyers had reached roughly $10 billion (Reuters records the characterization as "minimal"; TechTimes rendered it "trivial"), all follow the wire-service record. Huawei's Ascend 950PR was unveiled March 20, 2026, with mass production beginning the following month, per China AI Dispatch. China now accounts for less than 1 percent of Nvidia's data-center revenue, per Bloomberg's earnings-week analysis and Motley Fool's writeup; Huang's earlier $50 billion opportunity framing is from Fortune and CNBC's coverage of his May 2025 remarks. The House Foreign Affairs Committee's advancement of the AI OVERWATCH Act by a 42-2 vote on January 21, 2026 is documented in the committee's own press release, The Hill, and Bloomberg. The physical-AI revenue reading is the most exposed figure in the chapter: Kress said on the Q4 FY26 call, per The Globe and Mail's transcript, that physical AI had contributed "more than $6 billion" in fiscal-year 2026, and Morningstar tracked the number to roughly $9 billion on a trailing-twelve-month basis by mid-2026.

Chapter 15: Where the Bets Land

This chapter carries the heaviest citation load in the book, seventy-two verified claims, and its sourcing follows a tiering discipline: company filings, earnings calls, and official announcements first; dated, named byline reporting second; analyst estimates third, always attributed as estimates in the text.

The Microsoft material, the Frontier Company launch, the roughly 27 percent OpenAI stake, the first-half stock decline, and Judson Althoff's neutral integrator pitch, follows the company's announcements and the contemporaneous financial press; the FTC's inquiry into the Microsoft-OpenAI structure remained open as this book went to press. The Google material, the 10-K's "full-stack approach" language, the eighth-generation TPU line with roughly 4.3 million units projected to ship in 2026 (a semiconductor-analyst estimate, attributed as such), and the Gemini app passing 900 million monthly active users, follows Alphabet's filings and disclosures.

The forward-deployed-engineer convergence, OpenAI's and Anthropic's programs launching in May 2026 and Amazon's billion-dollar commitment on June 30, follows the companies' announcements. The Palantir-Nvidia partnership on classified and air-gapped deployments is characterized strictly in the terms the companies have used publicly. The Oracle credit picture is sourced under the Conclusion. The SpaceX and Colossus material is sourced under Chapter 7, and the Meta and Mistral material under Chapters 8 and 10. Where this chapter's callbacks repeat claims made earlier in the book, the earlier chapter's sourcing governs.

Chapter 16: The AI Industrial Complex

Eisenhower's farewell address was delivered January 17, 1961, three days before he left office; the "military-industrial complex" and "scientific-technological elite" passages are quoted verbatim from the National Archives text, as is the "government contract becomes virtually a substitute for intellectual curiosity" line. Biden's farewell address of January 15, 2025, with its "tech-industrial complex" warning, follows the Miller Center text and Reuters coverage; the inauguration seating five days later was documented across the *Guardian*, *Newsweek*, and the *Telegraph*.

Eric Schmidt was booed at the University of Arizona commencement on May 15, 2026, per the *Arizona Daily Star* and *The Verge*. The pattern of layoffs announced alongside record AI capital expenditure is sourced under Chapter 17, where it is treated in full.

The ocean and orbit material: the Shanghai Lingang undersea data center, wind-powered and seawater-cooled, entered full operation in May 2026, per the *Guardian*, *Tom's Hardware*, and the *South China Morning Post*; SpaceX's June 2026 FCC filing contemplating up to a million orbital compute satellites was reported by *USA Today*; Starlink's foreign policy weight in Ukraine follows *Foreign Policy*'s analysis and the Reuters investigation into service shutdowns.

The European consultation frameworks are the Collective Redundancies Directive (98/59/EC) and the European Works Councils Directive (2009/38/EC). Vail's coining of universal service is sourced under Chapter 4. The philanthropy comparisons, Carnegie's roughly 2,500 libraries, the Rockefeller institutions, the Gates Foundation's scale, follow the institutions' own histories. The Intel stake and the OpenAI equity offer are sourced under Chapters 9 and 13.

Chapter 17: Who Bears the Cost

The Spartanburg story, the Project MOC-1 codename, the officials avoiding the words "data center" for over a year, and David Britt's move to the Public Service Commission followed by the recusal motion against him, follows the *Washington Post*'s investigation (July 3, 2026) and Spartanburg County news coverage. The emissions figures come from South Carolina Department of Environmental Services permit filings; the Southern Environmental Law Center's assessment that the limits were written "loosely enough to be functionally unenforceable" is quoted from its public comments.

Anne Hedges's "easy pickings" is verbatim; she is the executive director of the Montana Environmental Information Center, where she has worked for more than thirty years. The Stratos thermal-load analysis follows the public work of Utah State University physicist Rob Davies.

The BlackRock-led acquisition of Aligned Data Centers was announced October 15, 2025; Larry Fink's statement about "mobilizing private capital to accelerate AI innovation" comes from the Global Infrastructure Partners release of November 14, 2025. The Abilene, Texas, project is the OpenAI-Oracle venture built on the Stargate commitment.

The layoff figures follow the major trackers: TechCrunch's layoff list and the Challenger, Gray year-end report, with broader trackers approaching a quarter million tech jobs cut in 2025. Marc Benioff's "I need less heads" remarks were made on *The Logan Bartlett Show* podcast in September 2025 and widely reported. The New Jersey data center tariff bill passed the legislature July 1, 2026, and was sent to Governor Sherrill's desk; as of this writing it awaits signature.

Chapter 18: The Strategic Capture

The market-concentration figures: Nvidia's roughly 7 to 8 percent weight in the S&P 500 is the highest single-stock concentration in the index's history, exceeding Exxon Mobil's roughly 4 to 5 percent peak in 2008, per the index-concentration analyses cited in the text. The comparison of data center construction spending to total US transportation infrastructure spending follows the 2026 construction spending data.

The IBM quantum foundry arrangement, roughly $1 billion tied to a $2 billion venture, follows the companies' announcements. The named commentary, Ivo Daalder and Jim Townsend (a former US assistant secretary of defense for Europe and NATO) on the Pentagon's six-month review of US force posture in Europe, Gary Marcus, and the analysts quoted on regulatory capture, is drawn from each figure's public statements and attributed reporting; where this chapter's editorial pass could not verify a direct quotation, the text paraphrases and attributes rather than quotes. The Sanders bill mechanism and the OpenAI lobbying material are sourced under Chapter 13.

Chapter 19: Safety Governance

The epigraph is Benjamin Franklin's from 1755.

A note on method for this chapter specifically: The litigation it describes is characterized strictly in the terms the public record itself uses, court filings, docket entries, and named-byline coverage, and no claim about any case goes beyond what those records state. Legislative statuses are stated as of mid-2026 and were verified at that date: the federal preemption provision received no floor vote; the Great American AI Act was a circulating discussion draft not yet in committee; Connecticut's dynamic pricing law was enacted while Nebraska's remained proposed; and the executive order language on federal funding conditions is paraphrased from the order's text rather than from characterizations of it. NSPM-7's wording is quoted in its own noun forms. Readers checking any of these against the record after mid-2026 should expect the statuses, though not the underlying documents, to have moved.

Conclusion: The Grid Beneath the Cathedral

The epigraph is Alfred North Whitehead's from *An Introduction to Mathematics* (1911).

The fiber build out accounting, roughly $90 billion for the core long-haul networks within a broader half-trillion-dollar telecom capital wave, and more than a trillion dollars of market value destroyed, follows the contemporaneous tallies, including the *Los Angeles Times* reckoning of June 2002. The 90 percent collapse in bandwidth prices follows *Wired*'s 2004 retrospective and the historical transit-pricing series maintained at DrPeering.

The current build-out's scale: $725 billion in 2026 hyperscaler capital expenditure, up 77 percent year over year, per the companies' own guidance; McKinsey's estimate of $6.7 trillion in data center investment by 2030 from "The Cost of Compute" (April 2025); the IEA's projection of 945 terawatt hours by 2030. The roughly $108 billion in 2025 hyperscaler bond issuance follows credit-market analyses, with counts ranging to $121 billion depending on methodology.

The Oracle figures: a capex-to-revenue ratio of 83 percent for fiscal 2026, per its earnings; negative outlooks from two of the three major agencies (Moody's and S&P, 2025), with Fitch affirming BBB with a stable outlook in February 2026; and credit-default-swap spreads widening roughly fivefold, from about forty basis points in early 2025 to nearly two hundred by March 2026. The group funding mix, roughly ninety cents of capex per dollar of operating cash flow by 2026, up from about sixty-five cents a year earlier, with Amazon's quarterly capex exceeding its operating cash flow, follows 2026 financial reporting.

Pope Leo XIV's first encyclical, *Magnifica Humanitas*, "On Safeguarding the Human Person in the Time of Artificial Intelligence," was signed May 15 and published May 25, 2026, and personally presented by the pope; my account of it, and of the clergy, academics, and ethicists the labs themselves have convened, draws on my own Techstrong.ai reporting ("Hello God, It's Me, Claude," April 15, 2026, and the Vatican piece of May 26, 2026), along with my July 2025 essay on the builders of cathedrals.

Daniel Newman's bull case is made across his LinkedIn commentary, his CNBC appearances, and the Techstrong piece this conclusion engages. Gennaro Cuofano's "AI & The Dynamo Doctrine" was published on *The Business Engineer* in June 2026; his substrate lineage (electrons into bits, bits into tokens) and his seventy-two-hour jurisdictional-shutoff observation are quoted from that essay.

The historical callbacks are sourced under their home chapters: Jevons's *The Coal Question* (1865) under Chapter 1, Insull's 1932 collapse under Chapter 5, the Kingsbury Commitment (1913) and the 1956 and 1982 consent decrees under Chapter 4. Paul David's dynamo argument is from his 1990 *American Economic Review* paper. The Spartanburg project was withdrawn on July 1, 2026, after county officials denied its tax arrangement, which is why the closing paragraph counts it among the places that said no.

Acknowledgments

Writing my first book—something I have threatened to do for twenty-plus years—does not get done without a lot of help. Not just from the people I work with now, but from all the people who helped a frustrated lawyer and closet computer nerd spend more than thirty years doing what he loves, and so never worked a day in all that time. The people who helped along the way are too many to mention, and some of them are no longer here on this earth, which makes me appreciate what they gave me all the more.

First, from my early tech career, my first partner at TriStar Web, Billy Nicholson. Friends since we were babies, but who knew we would end up internet gurus?

Next, two people who were true mentors, not only in tech but, more importantly, in life and business. I mentioned Brad Feld, but Brad's help and influence deserve far more than a mention. I am always grateful to him for setting me out on this journey. He taught me how to act like I belong here. Brad's mentor, Len Fassler, who taught me that "whether it's baking chocolate chip cookies or web hosting babe, you have to know how to make a dollar." And so much more than that about building a business.

From my Interliant years, through StillSecure, all the way to Techstrong: my friend Rajat Bhargava. Sometimes he felt like my little brother, sometimes he was the older, more mature one, but he is still probably the smartest person I have ever met.

Speaking of StillSecure through Techstrong and now Futurum, my friend Mitch Ashley has been my peer, confidant, partner, and adviser for twenty-five years. I can't thank him enough.

So many others have helped along the way: Mike Vizard and Charlene

O'Hanlon, who taught me the media and editorial business, and Parker Yates, the sales guy straight out of central casting, who has been at it with me for twenty years. And everyone who made Techstrong what it is today.

Now, as part of Futurum: first and foremost, my friend, partner, and the ultimate chief analyst, Daniel Newman. It is his utter bullishness on AI that drives me to think harder and go looking for the stories this book is full of. I can't wait to see the even greater things he has yet to accomplish.

A very special shout-out to my friend John "Botchagalupe" Willis. My one regret is that I didn't meet John sooner. We grew up so close to each other and led strangely parallel lives, but we didn't meet until the DevOps thing. His help, advice, and guidance on how to write a book—and all that goes into it—was invaluable.

Thank you to Leah Brown for her editing of this book. Every author should be lucky enough to have a Leah to help them.

And in what may still be a novelty on an acknowledgments page but won't be for long: Claude, ChatGPT, and Perplexity. I have argued with them, debated them, and thought through every sentence of this book with them—research, source-checking, cover and jacket design, and more. A book about the fortunes built on top of commoditizing intelligence was, it seems only honest to admit, itself built on top of it.

I already thanked my immediate family in the dedication, but to my extended family—many of whom have worked at Techstrong over the years—thank you for all the hard work.

And to Kai'a, my chubby blue merle bulldog. More of this book than I'd like to admit was worked out on our walks, with her waddling along beside me while I argued with railroad barons in my head. Every author needs one source of unconditional love, and she is mine.

Any errors that remain are mine alone.

www.ingramcontent.com/pod-product-compliance
Lightning Source LLC
LaVergne TN
LVHW100725061126
851857LV00011B/407